CYBERACTIVISM

CYBERACTIVISM

Online Activism in Theory and Practice

EDITED BY

Martha McCaughey

AND

Michael D. Ayers

ROUTLEDGE
NEW YORK & LONDON

Published in 2003 by
Routledge
29 West 35th Street
New York, NY 10001
www.routledge-ny.com

Published in Great Britain by
Routledge
11 New Fetter Lane
London EC4P 4EE
www.routledge.co.uk

Printed in the United States of America on acid-free paper.

10 9 8 7 6 5 4 3 2 1

Library of Congress Cataloging-in-Publication Data

Cyberactivism : online activism in theory and practice / edited by Martha McCaughey and Michael D. Ayers.
p. cm.
Includes bibliographical references and index.
ISBN 0-415-94319-1 (alk. paper) — ISBN 0-415-94320-5 (pbk. : alk. paper)
1. Internet—Social aspects. 2. Internet—Political aspects. 3. Social movements.
4. Protest movements. 5. Social action. I. McCaughey, Martha, 1966– II. Ayers, Michael D.

HM851 .C93 2003
303.48'33—dc21 2002031834

CONTENTS

ACKNOWLEDGMENTS

We wish to thank Matthew Byrnie at Routledge, Megan Boler at Virginia Tech, Richard Widick at UCSB, and Christina French, all of whom offered helpful feedback on this project. Thanks especially to David Silver for his unending enthusiasm for cyberculture studies and his support for this project specifically.

INTRODUCTION

Martha McCaughey and Michael D. Ayers

The increasing commercialization of the Internet demands a scholarly and political response. A huge portion of the books about the Internet published in the last three years claim to teach how to make money using it. Harder to notice but important to explore is the presence of political activism on the Internet, which we are calling *cyberactivism.* Amidst the publication of *The Neatest Little Guide to Making Money Online* and *Starting an Online Business for Dummies®*, small and large networks of wired activists have been creating online petitions, developing public awareness Web sites connected to traditional political organizations (e.g., Amnesty International online), building spoof sites that make political points (such as worldbunk.org), creating online sites that support and propel real-life (RL) protest (e.g., a16.org, which stands for April 16, the date of the World Trade Organization [WTO] protest in Washington, DC), designing Web sites to offer citizens information about toxic waste, and creating online organizations (e.g., Indymedia.org) that have expanded to do traditional RL activities. Activists have not only incorporated the Internet into their repertoire but also, as this volume shows, have changed substantially what counts as activism, what counts as community, collective identity, democratic space, and political strategy. And online activists challenge us to think about how cyberspace is meant to be used.

We assembled *Cyberactivism* to document and discuss these significant political efforts. We did so not because we think the Internet is the latest technology for reaching the American Dream. Neither we nor the contributors to this volume are technological utopians or dystopians. At the same time, all share a modest belief that new technologies can become agents of progressive social change.[1] If the Internet can assist people seeking progressive social change, it will do so not as an inevitability, nor as a cause, but rather as a means of change alongside other forces. This book documents the struggles of those activists and intellectuals interested in the use of Internet technologies for a democratic politics.

Cyberactivism crosses disciplines, mixes theories with practical activist approaches, and represents a broad range of online activist strategies, from online awareness campaigns to Internet-transmitted laser-projected messaging. The essays in this collection highlight the importance of current social-movement theory, cultural studies, media studies, and cyberculture studies to explore how cyberactivism helps, hinders, and transforms social-change work. The contributors to *Cyberactivism* explore the tensions between commerce and community, information and empowerment, awareness and action, identity and change, democracy and delinquency, privacy and piracy, cyberprotest and security breaches, online bodies and meat bodies.

The interdisciplinary field of cyberculture studies offers many helpful starting points for this study of online activism. Brenda Danet's *Cyberpl@y: Communicating Online* (2001) shows how the Internet is transforming writing, performance, and art. Nancy Baym's study *Tune In, Log Out: Soaps, Fandom and On-line Community* (1999) and Sherry Turkle's *Life on the Screen* (1997) investigate identity and community on the Internet, but not as part of a political movement. The compelling edited collections *Race in Cyberspace* (Kolko, Nakamura, and Rodman 2000) and *Technicolor* (Nelson, Tu, and Hines 2001) show us the many forms of racism that continue in this new communications medium. *Cyberactivism* highlights the activism in this medium to challenge such injustices.

While the Religious Right, the Ku Klux Klan (KKK), and other radically conservative political organizations have also colonized cyberspace in hopes of achieving their goals (see, e.g., Tara McPherson's study of U.S. Southern nationalists [or neo-Confederates] online in Kolko, Nakamura, and Rodman 2000 and Susan Zickmund's study of radical white supremacists in Jones 1997), we chose topics that focus on social-change efforts that could loosely be called progressive or anti-establishment. Under this category we include single-incident or single-campaign political actions by substate actors known as hacktivism, but not hacking into databases to steal credit card numbers or the online actions by one government's military of breaking into another military computer system, sometimes referred to as cyberwar, cyberattack, or cyberterrorism. Attempting to understand political movements online urges the field of cyberculture studies to shift, as McPherson (2000, 12) noted, from a focus on play, mulitiplicity, rhetoric, and theater toward explorations of struggles, citizens, politics, and publics.

Previous studies in the field have taken advantage of the easy access the Web provides a university researcher—so, for example, a left-wing scholar who wouldn't get into a KKK meeting or rally can lurk in an online discussion list and record the postings. While such studies offer insights into the political sentiments of specific groups, this volume moves beyond documenting political views and asks how the Internet as an interface affects political sentiment, organizing, and social-movement theory. In other words, previous studies, while helpful for many reasons, have not studied online activism qua activism or qua cyberculture.

Understanding cyberactivism is important not only for scholars in cyberculture studies but also for scholars interested in activism, social transformation, and technology. Online activism raises new questions about political organizing and social change. Along with documenting new forms of activism enabled by the Internet, this volume engages social-movement theories about organizational structures, collective identity, leadership, and the framing of political issues. This book forms an interdisciplinary collection with critical examinations of a wide range of movements.

A broad spectrum of activists, whether or not they are (as yet) online, must assess the relevance of computer-mediated communication for their goals, constituencies, and opposition. It's a well-worn stereotype that activists rarely read about activism. After all, they're too busy doing "real" work to read ivory tower–generated accounts, critiques, and histories of social change and social-change work. But reflecting on, theorizing, and historicizing activism is crucial, and this volume puts activism in the perspective of broader trends in cyberculture, political activism, and social-movement theory.

Technology is hardly new to activists. Social-movement groups have historically incorporated new technologies into their social-change struggles. Whether newspaper, radio, TV, or film, activists have embraced new communications media to circulate information, make statements, raise consciousnesses, raise hell. Over a decade ago, Chinese student activists in the Tiananmen Square democracy movement used computer bulletin boards, and the peasants in Chiapis, Mexico, broadcast their struggles using computer databases in addition to guerrilla radio and other forms of media (Kellner 1997). But activists have simultaneously relied on in-the-flesh meetings (now called "meatings" by those frequently online) to make plans, form a collective identity, protest, or inspire a crowd.

In his account of the civil rights movement, Aldon Morris (1984) details the nonviolent tactics of the participants to garner support from the general public. We know some of those tactics thanks to TV—it's hard to erase stunning visual images of peaceful black demonstrators getting firehosed by whites determined to keep the status quo. It's equally difficult to forget the young student who stood before a government tank in the 1989 Tiananmen Square demonstration. Can the Internet play a role similar to the TV in creating a visual/textual/aural image in the public imagination? Is the Internet the same kind of media technology, with the same connection between group imagination and real-world bodily action?

How is the Internet different from previous communications media that have influenced the nature and shape of political organizing? The Internet is immediate, even more immediate than a daily newspaper. It can be more interactive than TV. Although the digital divide still keeps many of the poorest people and the poorest nations from access, and the Internet is still predominantly in English, it is increasingly accessible. It is not only instant and transspatial but multilateral, including many participants and connecting many different activist groups. Not since the institutionalization of the U.S. Postal Service have we seen a communication development in society that can give power to individuals like this.

The Internet allows us to interact with others without our voices, faces, and bodies. In the absence of meat bodies (also known as wetware among Netizens), the traversing of spatial and temporal boundaries raises questions about what presence, essence, or soul we think we are on the Net (Slater 1998). The Internet thus raises new questions about social change and how it works. For instance, where is the body on which that traditional activism has relied? Indeed, a substantial component of political activism has always been the activist's willingness to put her body on the line. After all, forwarding an online petition to your e-mail recipient list is far from sitting in at Woolworth's lunch counter or chaining yourself to an old tree. Can you really put your body on the line online? If not, how do social-change efforts online actually work? If politics takes place in pixels, when is the wetware necessary? In what ways does cyberspace activism demand, facilitate, or depend on meatspace or RL activism, and when can a purely textual presence actually cause political change?

Powerful government, corporate, and cultural organizations have traditionally understood the extent of opposition to them because people picketed, protested, wrote newspaper articles, and so on. Now, in addition and sometimes instead, political opposition gets organized, and expressed, online. When the World Bank/International Monetary Fund (IMF) protestors crowded the streets of Seattle in 1999, the associated Web sites and online multimedia projects that generated support for them could in no way have replaced the actual bodies on the streets. So how do we distinguish types of activism? Can the Internet be used *for* protest, or does it simply *support* RL protests?

As contributors to this volume show, new forms of protest do occur online, and these new forms raise new issues for activists and the scholars studying activism and cyberspace. For example, hacktivists conduct online sit-ins and deface Web pages as a form of protest. Is such hacktivism going to be framed as a national security breach, as a case of bored hackers using politics as their pretext, or as a necessary, new form of political direct action? How do we assess the types of activism on the Internet today, and more vexing, how do we assess the ethics and effectiveness of this work? After the September 11, 2001, terrorist attacks in New York City and Washington, DC, some people felt politically active for the first time—because they were on the Internet. One woman, featured in a newspaper article (Yim 2001), had gone from voting in the big elections and not otherwise being politically involved to saying, "I feel like a political activist, for the first time in my life. . . . If enough people look at a peaceful way to resolve all the world conflict, it's pretty powerful. . . . The e-mail is the best thing we've got going. The Internet is powerful" (C1, C3). The events of September 11 acted as the catalyst for this person's activism, making her feel active through the Internet thanks to its convenience. But has an activist really been born? Is texting tantamount to protesting?

Much of what seems to pass for political work online involves the circulation of electronic legends: warning one's friends and family not to go to a shopping mall on Halloween because the friend of a friend's boyfriend might have been an Al Qaeda terrorist who issued that warning, or forwarding the infamous e-mails from the Brandeis University student, in circulation since 1999, about the position of women in Afghanistan.[2] People feel like they are doing something useful when they push the send button, even though spamlike petitions and e-mailed chain letters have not been proven effective and are often experienced by recipients and systems administrators as unethical abuses of network systems. If we don't even know who gets the e-mail petition that we've just e-signed, then what counts as political involvement? Some "real" activists may even worry that political apathy grows with each new Internet connection. But those activists are also using the Internet in droves.

Online activism raises questions for social-movement scholars as well. Academic theories that explain the success of social movements suggest that social-change action occurs because of political opportunities that lie within the state, because strong collective identities exist within groups and movements, and because a social-movement group has the ability to "frame" its cause successfully. Since the 1970s, theorists of social movements (e.g., McCarthy and Zald [1973, 1977; Oberschall 1973; and Tilly 1978) explained a social movement's success through a theory of resource mobilization. These scholars argued that a movement's success can be judged by looking at whether resources (e.g., money, media access with a favorable depiction of the cause) are readily available to the movement group. The grievances a group might have with society are largely discredited in this model because grievances alone have not, historically, created social movements (Buechler 2000). A better predictor of movement success would be the ability for a group to gain resources that it could use in order to garner support.

An extension of resource mobilization was developed by Doug McAdam (1982), who looked at politics and political opportunities that the state might be fostering (usually unintentionally) or not fostering as a predictor of a movement's success. In *Power in Movement: Social Movements and Contentious Politics* (1998), Sidney Tarrow provides an example of how political opportunities can allow for movements to make an impact. In the 1930s, the depression that spread through Western Europe and the United States led to movements as a result of political opportunities. Workers in Britain and Germany did not have the opportunity to create movements based on their work and economic conditions, yet in France and the United States (countries enduring the same economic depression) workers found ways to strike and create successful movements. Political reforms that occurred in France in 1933 (the French Popular Front) and in the United States in 1936 (the New Deal) created an atmosphere that was reluctant to suppress labor and, in turn, created an innovative economic-political market that could directly impact workers. Thus, with this example, Tarrow concludes that labor insurgency was possible because of the new opportunities created by the state, and not worker grievances or their available resources (Tarrow 1998, 73). So, a main argument of the political-opportunities theory examines state strength, prevailing strategies, and state repressiveness (Tarrow 1998). Movements that are centered around class, labor, and economics can use this model to explain the movements' emergence.

While the political-opportunities model shows how a movement can emerge, it does not provide much insight into how a movement can sustain itself over time. In the late 1980s, social-movement scholars such as Alberto Melucci (1989) and Verta Taylor and Nancy Whittier (1992) started to examine this issue and research a social-movement group's collective identity. Collective identity is a social psychological concept that explains the link an individual has to the group. When a group has a strong collective identity, the movement can garner support and power because the participants feel that they are all working toward common goals, have defined opponents, and have an integrated sense of being that is incorporated into the movement ideologies. Identity-based movements, as opposed to class-based movements, have been central to social-movement research in the past decade, as a result of the gay and lesbian movements, the women's movement, the civil rights movement, and some aspects of the alternative-globalization movement.

In terms of political opportunities, we can raise a question: To what extent does the Internet create or not create activist opportunities? The Internet began as a government-funded project in the late 1960s, but it has been mainstreamed into the general public for use and consumption by whomever can obtain access. (See Kidd in this volume for a historical view of the Internet's development.) Chapters in this volume thematically pose the following question: Has the Internet become a new political opportunity for activists because it is, in many places, no longer under strict government control and development? This volume also asks whether collective identity is possible over the Internet, where groups are dispersed across time and space and gather only in a virtual space. Or does online activism show us that collective identity is not necessary for collective action for social change? Do groups need in-the-flesh contact for a collective identity to be generated? A group with a collective identity shares common concerns, a common enemy, and, typically, a common space. What does a community feel and look like online? How does it get organized and defined? How do we distinguish commercial appropriations of community from collective identity on the Net?

Several chapters in this volume examine Web sites and deconstruct their tactics for forming and framing its group members. Are these Web sites framing their message in an effective way? Are recruitment strategies being incorporated into social-movement groups' Web sites effectively? How do we distinguish commercial appropriations of community from collective identity that will lead to social change? Collective identities may be solidified online only to be appropriated by capitalist interests that many online community members may take for more empowerment/collective identity. Anthony Giddens (1991, 214) notes the significance of life politics—a politics not of emancipation but of life decisions, self-identity, and lifestyle—but warns against the commercialization of identities and lifestyles. Lifestyles can be packaged and sold to us as commodities and thus cannot simply represent freedom, democracy, and empowerment. How can we distinguish commercialization from empowerment groups? Concerns about the commercialization of the Internet—and, more broadly, technology's role in spreading global corporate capitalism—run through *Cyberactivism.* Contributors ask not only what globalization feels like online and how it can be resisted online but also how corporate interests appropriate the democratic dialogue and diversity of specific groups organized around lifestyles.

Many people take online quizzes for fun and pass them on to friends through e-mail (see, e.g., thespark.com). Taking and sharing the "How Gay Are You?" quiz[3] might allow the building of an identity-based group that, in the case of marginalized sexual identities, can be empowering in itself. But these online identity and lifestyle quizzes are thinly veiled marketing schemes. In addition to asking questions about sexual desires and identifications, test takers are also asked what type of products they like (do you wear boxer shorts, bikinis, or thongs?) and their age, zip code, and e-mail address. Some of these quizzes even find a way to tie in what kind of cereal you like for breakfast.

Market researchers and corporations selling items have used the Internet brilliantly—they have figured out ways people will joyously, enthusiastically give information about who they are, what they like, and where they can be reached. E-advertisements are now being made to look like cool lifestyle Web sites. A popular Web site among teenagers in 2000 and 2001, the "Curry" site (web.archive.org/web/20000815064734/http://www.rubberburner.com), appears to be the homepage of a young man named Curry who has a quirky personality. This site, which teenagers made one another aware of through their online peer-to-peer networks, was actually a corporate-produced advertisement for jeans (Harrison 2001; Kahney 2000; Silver 2001). The commercialization of the Internet has meant the employment of any means necessary for getting marketing and market research accomplished, thereby blurring lines between empowerment and commodification, freedom and constraint, having fun and being a stooge.

Such viral marketing schemes over e-mail are also being used by activists. In 2001, then MIT graduate student Jonah Peretti called attention to corporate abuse of laborers when millions of people received the e-mail exchange he had with Nike documenting the company's refusal to customize his personalizable shoes with the word "sweatshop" (Peretti 2001). That Peretti intended to buy a Nike product with the ironic label of "sweatshop" on it was itself a strategy of culture jamming, that is, using corporate power against itself through co-opting, hacking, ironic humor, or recontextualizing meanings (Peretti 2001). Peretti placed his order for shoes marked with "sweatshop" on the Nike Web site, and the exchange began. Peretti compiled the e-mail exchange into one message and sent it to twelve people. This is the compilation of those e-mails, which traveled across the Net:

> From: "Personalize, NIKE iD" <nikeid_personalize@nike.com>
> To: "'Jonah H. Peretti'" <peretti@media.mit.edu>
> Subject: RE: Your NIKE iD order o16468000
>
> Your NIKE iD order was cancelled for one or more of the following reasons.
> 1) Your Personal iD contains another party's trademark or other intellectual property.
> 2) Your Personal iD contains the name of an athlete or team we do not have the legal right to use.
> 3) Your Personal iD was left blank. Did you not want any personalization?
> 4) Your Personal iD contains profanity or inappropriate slang, and besides, your mother would slap us.
> If you wish to reorder your NIKE iD product with a new personaliza-

tion please visit us again at www.nike.com
Thank you,
NIKE iD

From: "Jonah H. Peretti" <peretti@media.mit.edu>
To: "Personalize, NIKE iD" <nikeid_personalize@nike.com>
Subject: RE: Your NIKE iD order o16468000

Greetings,
My order was canceled but my personal NIKE iD does not violate any of the criteria outlined in your message. The Personal iD on my custom ZOOM XC USA running shoes was the word "sweatshop." Sweatshop is not: 1) another's party's trademark, 2) the name of an athlete, 3) blank, or 4) profanity. I choose the iD because I wanted to remember the toil and labor of the children that made my shoes. Could you please ship them to me immediately.
Thanks and Happy New Year,
Jonah Peretti

From: "Personalize, NIKE iD" <nikeid_personalize@nike.com>
To: "'Jonah H. Peretti'" <peretti@media.mit.edu>
Subject: RE: Your NIKE iD order o16468000

Dear NIKE iD Customer,
Your NIKE iD order was cancelled because the iD you have chosen contains, as stated in the previous email correspondence, "inappropriate slang".

If you wish to reorder your NIKE iD product with a new personalization please visit us again at www.nike.com
Thank you,
NIKE iD

From: "Jonah H. Peretti" <peretti@media.mit.edu>
To: "Personalize, NIKE iD" <nikeid_personalize@nike.com>
Subject: RE: Your NIKE iD order o16468000

Dear NIKE iD,
Thank you for your quick response to my inquiry about my custom ZOOM XC USA running shoes. Although I commend you for your prompt customer service, I disagree with the claim that my personal iD was inappropriate slang. After consulting Webster's Dictionary, I discovered that "sweatshop" is in fact part of standard English, and not slang. The word means: "a shop or factory in which workers are employed for long hours at low wages and under unhealthy conditions" and its origin dates from 1892. So my personal iD does meet the criteria detailed in your first e-mail.

Your web site advertises that the NIKE iD program is "about freedom to choose and freedom to express who you are." I share Nike's love of freedom and personal expression. The site also says that "If you want it done right. . . . build it yourself." I was thrilled to be able to build my own shoes, and my personal iD was offered as a small token of appreciation for the sweatshop workers poised to help me realize my vision. I hope that you will value my freedom of expression and reconsider your decision to reject my order.

Thank you,
Jonah Peretti

From: "Personalize, NIKE iD" <nikeid_personalize@nike.com>
To: "'Jonah H. Peretti'" <peretti@media.mit.edu>
Subject: RE: Your NIKE iD order o16468000

Dear NIKE iD Customer,
Regarding the rules for personalization it also states on the NIKE iD web site that "Nike reserves the right to cancel any Personal iD up to 24 hours after it has been submitted".

In addition it further explains:
"While we honor most personal iDs, we cannot honor every one. Some may be (or contain) others' trademarks, or the names of certain professional sports teams, athletes or celebrities that Nike

does not have the right to use. Others may contain material that we consider inappropriate or simply do not want to place on our products.

Unfortunately, at times this obliges us to decline personal iDs that may otherwise seem unobjectionable. In any event, we will let you know if we decline your personal iD, and we will offer you the chance to submit another."

With these rules in mind we cannot accept your order as submitted.

If you wish to reorder your NIKE iD product with a new personalization please visit us again at www.nike.com

Thank you, NIKE iD

From: "Jonah H. Peretti" <peretti@media.mit.edu>
To: "Personalize, NIKE iD" <nikeid_personalize@nike.com>
Subject: RE: Your NIKE iD order o16468000

Dear NIKE iD,
Thank you for the time and energy you have spent on my request. I have decided to order the shoes with a different iD, but I would like to make one small request. Could you please send me a color snapshot of the ten-year-old Vietnamese girl who makes my shoes?
Thanks,
Jonah Peretti

{no response}

That the letters between Peretti and Nike were both amusing and politically significant helps explain the Nike e-mail's infective power. The Nike e-mail had spread like a virus to the computer screens of millions around the world, across a wide range of online social networks. The Nike sweatshop controversy was covered in newspapers, magazines, and on NBC's *Today* show, on which Peretti appeared, underscoring that cyberactivist strategies can harness long-standing forms of mass media communication. The culture-jamming e-mail turned the corporate logic of viral e-mail marketing against itself, a strategy other activists have adopted since. Internet technologies can be used for corporate manipulation, profit, domination, liberation, empowerment, enlightenment, and political resistance. This volume stresses the role progressive activists and public intellectuals can play in shaping new technologies.

In so doing, *Cyberactivism* refuses to define the boundaries of online activism or to determine what counts and does not count as legitimate online activism. (But see Vegh in this volume for a useful classification of online activist forms.) Defining online activism is as difficult as defining activism before the Internet. Activism takes many forms—including direct action, protests, efforts to change laws, self-help groups, educational groups, cultural groups, activist newspapers, and political bookstores. Activists and scholars have long debated the legitimacy of various forms of social-change efforts. In the feminist movement, for instance, those who run a feminist bookstore might be seen as too conservative by those struggling to pass the Equal Rights Amendment, who might be seen as too conservative by those who march in front of a porn store, who might be seen as too conservative by those who deface a billboard objectifying women, who might be seen as too conservative by feminists who formed their own porn-production company, and so on. Specific activist strategies may be symbolic but may not lead to any individual or structural change. For instance, the well-known Take Back the Night March and Rally is a symbolic gesture; we know that it does not lead directly to the arrest of rapists or to a reduction in the rape rate. But most would probably agree that it (at least in its less institutionalized days) counted as a bit of activism. In this same vein, we might ask, did the viral e-mail between Jonah Peretti and Nike cause Nike to increase its overseas employees' wages? If not, does it still count as activism? This book cannot decide whether or not culture jamming should or should not count as activism, but its study of activism in the new context of cyberspace certainly focuses attention on these very questions.

Looking Ahead

We have divided *Cyberactivism* into three sections. In part I, Cyber-Social Movements Emerging Online, contributors show us new social movements that have emerged as a direct result of Internet technologies. Activists often utilize the Web to recruit, strategize, and create change, and some activism fixes on the politics of the Net itself.

In chapter 1, Laura J. Gurak and John Logie trace two Web-based protests *around* the Internet. This chapter reveals the Internet to be a space for activists not only to promote and rally around RL movements, but also to challenge injustices that occur on the Internet itself. Presenting cases of Lotus MarketPlace and the Clipper chip, the more recent Web-based protests enabled by sites such as Petitionsite.com, and the protest against Yahoo!'s attempt to appropriate Web spaces built by "citizens" of GeoCities, Gurak and Logie analyze the emergence, because of the Internet itself, of new movements that protest the corporatization of the Internet.

In chapter 2, Dorothy Kidd examines the trials and tribulations of launching a social-change group on the Internet, the Independent Media Center. In so doing, she discusses the kind of public space the Internet makes possible for an alternative media source devoted to social change. Further, by tracing the Seattle organizers' use of the Internet and its multimedia capacities to bypass corporate-controlled media, Kidd shows powerful ways in which online and RL anti-corporate globalization efforts support one another.

In chapter 3, Sandor Vegh classifies the many forms of online activism, offering useful ways of distinguishing types of cyberactivism, including hacktivism. From virtual sit-ins to Web site defacements, Vegh examines hacktivist techniques against the state and big business in the anti-globalization movement, presenting hacktivism as a new form of protest unique to cyberspace. This research raises questions about parallels between hacktivism and RL activism. Is hacking into the online World Bank meetings with e-mail spams, for instance, the equivalent of defacing a building or planting a bomb in a building, or is hacktivism a newer, less violent form of protest unique to the Internet, demanding different ethical considerations? How do the politically powerful frame and respond to hacktivism? Do they consider it a dangerous crime because it is harmful and illegal, or simply because it threatens their control?

In chapter 4, Larry Elin presents a case history of Zeke Spier, a college student activist involved in the anti-globalization movement. Zeke Spier combined tactics of traditional activism (e.g., jail solidarity) with online information dispersal and gathering. Through this case study, Elin reveals the Internet to be a powerful recruitment tool, having ultimately led to Zeke Spier's dropping out of Brown University and joining a protest in Philadelphia with people he met through the Internet. With this case study, Elin reveals the Internet as a transformative political force in a person's life, and the extent to which online activism blended with RL activism.

In part II of the collection, Theorizing Online Activism, contributors take theoretical approaches that have been applied to traditional social movements and examine how well they stand the test of cyberspace. Lee Salter begins this section by taking Jurgen Habermas's theory of the public sphere and the lifeworld to assess how new social movements are incorporating these concepts in their self-understandings and missions on the Internet. Salter looks at the Association for Progressive Communicators (APC) to analyze how this group attempts to create a network of nongovernmental organizations (NGOs) to support individuals and social movements in participatory political processes. Using Habermasian theory Salter makes sense of the Internet as a decentered textual communications medium used by new social movements for nonhierarchical communication, while also revealing that cyberspace, and the APC in particular, is not the ideal, democratic speech situation.

In chapter 6, Michael D. Ayers presents a comparative study of two feminist activist groups: one online and another offline. Through interviews with both sets of activists, he explores the notion of collective identity to discern whether an online activist group can create the same strong collective identity that effective groups typically experience when they have face-to-face interaction. Ayers found that online feminist activists talked mainly about meeting and having sex with other activists, while the offline group got some important work done. Since social-movement theory presumes the necessity of a collective identity for social change, Ayers's study raises the possibility that online collective identities either do not always exist among online activists or do not necessarily lead to social-change efforts.

In chapter 7, Maria Garrido and Alexander Halavais apply a social-networks analysis in their study of the Zapatista movement. The Zapatistas are known as a social movement of the information age, constituted in and of the global communications network. Garrido and Halavais describe how these networks are used by mapping hyperlinks between sites to reveal the greatest activity "regions." Their research method reveals the extent to which the Zapatistas interact with a larger network of NGOs and proves useful for future research on how social-movement groups network.

In chapter 8, Wyatt Galusky explores the anti-toxins movement online, a social movement that falls under the environmental activism umbrella. Galusky examines Scorecard.org, a Web site devoted to the anti-toxins movement, calling into question the level of activism this site can promote. By showing the limitations of information provided online and how activists are discursively positioned in relationship to that information, Galusky raises important questions about how activist Web sites frame the citizen, the expert, information, and empowerment.

Part III of the collection, Cautionary Readings of Community, Empowerment, and Capitalism Online, examines ways in which traditionally RL social movements are incorporating the Internet into their activist repertoire and provides some cautionary tales about the commercialization of this process. In chapter 9, Joanne Lebert, an Amnesty International member and human rights activist, critically examines her own organization's use of the Internet. She calls into question the ability of Amnesty to use the Web for effective public outreach, as well as the capacity of those outside the organization to create false Web sites under the Amnesty name, thus raising the issue of false representation on the Web.

In chapter 10, Steven McLaine examines three ethnic online communities, paying particular attention to the corporate tactics used to create an empowered identity as a racial/ethnic community online. Studying AsianAvenue.com, BlackPlanet.com, and MiGente.com,

McLaine reveals a strong tension between online community and the capitalist interests driving the creation of the major racial/ethnic group Web sites. McLaine's chapter warns us against assuming that racial/ethnic unity online necessarily creates a social movement for social justice.

Part III concludes with Joshua Gamson's discussion of Gay Media, Inc. Gamson shows that Web sites devoted to the lesbian, gay, bisexual, and transgender community are commercialized entities, and as such champion diversity while being quite uniform and closing off the critical exchange of ideas historically so strong in the gay rights movement. Like McLaine, Gamson raises questions concerning a social-movement conglomerate on the Web, pushing us to think carefully about collective identities for corporate interests of capitalist consumption versus collective identities for empowerment and social justice.

Our volume concludes with an epilogue by the founder of the Resource Center for Cyberculture Studies,[4] David Silver, who situates this examination of online activism in the field of cyberculture studies. In so doing, he announces important new horizons for both academic work and activist history.

Notes

1. Our own forays into online activism can be summarized as follows.

Martha: I learned the political usefulness of publishing a Web page when my university's newspaper refused to print a statement, "racism@vt.edu," that some of my colleagues and I wrote protesting the way my university handled a racist e-mail message that was sent to a university listserv. We used the university server space (which each of us was granted when we got a university e-mail address) to post the statement and then broadcast its publication widely. I also started teaching online feminist courses because of a variety of institutional changes (see my explanation of how "football drove us to cyberspace" in McCaughey and Burger 1999). Since women's studies embraces the classroom as a political, consciousness-raising space and as a potentially empowering space for women, I concerned myself with what kind of feminist activism I could do teaching in cyberspace. I also studied the extent to which men and women had a political experience in a women's studies online class. Most recently, after *Cyberactivism* was already underway, I (along with several others) received an anonymous e-mail spam containing the political manifesto of a group claiming responsibility for spraypainting political graffiti on my campus. The campus police seized my computer without a warrant, claiming that since the university provides professors with the computers, they can take the computers and the files stored on them, whenever they want to. Thus I developed a new level of interest in cyberliberties in an increasingly computerized workplace.

Michael: As a fan of the band Phish, I noticed other fans at the concerts wearing T-shirts denoting their *online* fan identity, for example "Phish.Net" and "Phunky Bitches" (www.phunky.com). (For a study of Phish.Netters, see Watson 1997.) Such cybersubcultures within a music subculture made me wonder why people mobilized through the Internet around something as mundane as a band, and if something similar was happening in the political realm, where the stakes are more obvious. I was curious about my own online apathy. After all, I never responded to the e-mail petitions I'd been receiving monthly ever since I first got an e-mail account. I became curious how the bonds of community formed by Phish fans online might actually be forged among political movements for successful organizational effect.

2. For that petition, see http://urbanlegends.about.com/library/blafghan.htm. For other online petitions like it and for more information about urban legends that circulate over the Internet, see http://urbanlegends.about.com and http://vmyths.com/.

3. For that quiz, see http://test3.thespark.com/gaytest/. For other identity and personality quizes, see http://www.thespark.com.

4. For the Resource Center for Cyberculture Studies, see http://www.com.washington.edu/rccs.

References

Baym, Nancy. 1999. *Tune In, Log Out: Soaps, Fandom and On-line Community*. Thousand Oaks, CA: Sage.

Buechler, Steven M. 2000. *Social Movements in Advanced Capitalism.* New York: Oxford University Press.

Danet, Brenda. 2001. *Cyberpl@y: Communicating Online*. New York: Berg.

Giddens, Anthony. 1991. *Modernity and Self-Identity*. Cambridge, England: Polity Press.

Harrison, Amy. 2001. "Where Are They Now? Online Identities on the Commercial Web." Master's Thesis. Communication, Culture and Technology, Georgetown University.

Jones, Steven G. 1997. *Virtual Culture: Identity and Communication in Cybersociety*. Thousand Oaks, CA: Sage.

Kahney, Leander. 2000. "Shades of Mahir? Just Slight-Lee." *Wired News*. August 3. Online. http://www.wired.com/news/print/0,1294,37985,00.html.

Kellner, Douglas. 1997. "Intellectuals, the New Public Spheres, and Techno-Politics." Online. http://www.gseis.ucla.edu/courses/ed253a/newDK/intell.htm.

Kolko, Beth E., Lisa Nakamura, and Gilbert B. Rodman, eds. 2000. *Race in Cyberspace*. New York: Routledge.

McAdam, Doug. 1988. *Freedom Summer*. New York: Oxford University Press.

McCarthy, John D., and Mayer N. Zald. 1973. *The Trend of Social Movements in America*. Morristown, NJ: General Learning Press.

——. 1977. "Resource Mobilization and Social Movements." *American Journal of Sociology* 82: 1212–41.

McCaughey, Martha, and Carol J. Burger. 1999. "Cybergrrrl Education and Virtual Feminism: Using the Internet to Teach Introductory Women's Studies," pp. 151–61 in *Teaching Introduction to Women's Studies: Expectations and Strategies*, ed. Barbara Scott Winkler and Carolyn DiPalma.

McPherson, Tara. 2000. "I'll Take My Stand in Dixie-Net: White Guys, the South, and Cyberspace," pp. 117–31 in *Race in Cyberspace,* ed. Beth E. Kolko, Lisa Nakamura, and Gilbert B. Rodman. New York: Routledge.

Melucci, Alberto. 1989. *Nomads of the Present: Social Movements and Individual Needs in Contemporary Society*. Philadelphia: Temple University Press.

Morris, Aldon D. 1984. *The Origins of the Civil Rights Movement*. New York: The Free Press.

Nelson, Alondra, Thuy Linh N. Tu, and Alicia Headlam Hines, eds. 2001. *Technicolor*. New York: New York University Press.

Oberschall, Anthony. 1973. *Social Conflict and Social Movements*. Englewood Cliffs, NJ: Prentice-Hall.

Peretti, Jonah. 2001, April 9. "My Nike Media Adventure." *The Nation*. www.thenation.com/doc.mhtml?i=20010409&s=peretti.

Silver, David. 2001, April 1. "Discourses of Cyberspace: Culture, Community, Consumption." Presentation at the Institute for the Social Assessment of Information Technology, Virginia Tech, Blacksburg, Virginia.

Slater, Don. 1998. "Trading Sexpics on IRC." *Body and Society* 4 (4):91–117.

Tarrow, Sidney. 1998. *Power in Movement: Social Movements and Contentious Politics*. Cambridge, England: Cambridge University Press.

Taylor, Verta, and Whittier, Nancy. 1992. "Collective Identity in Social Movement Communities: Lesbian Feminist Mobilization," pp. 53–76 in *Frontiers in Social Movement Theory*, ed. Aldon D. Morris and Carol McClurg Mueller. New Haven, CT: Yale University Press.

Tilly, Charles. 1978. *From Mobilization to Revolution*. Reading, MA: Addison-Wesley.

Turkle, Sherry. 1997. *Life on the Screen*. New York: Touchstone Books.

Watson, Nessim. 1997. "Why We Argue About Virtual Community: A Case Study of the Phish.Net Fan Community," pp. 102–32 in *Virtual Culture: Identity and Communication in Cybersociety,* ed. Steven G. Jones. Thousand Oaks, CA: Sage.

Yim, Su-Jin. 2001, September 29. "Terrorist Attacks Cause Surge of Cyberactivism." *The Oregonian*, C1, C3.

Zickmund, Susan. 1997. "Approaching the Radical Other: The Discursive Culture of Cyberhate," pp. 185–205 in *Virtual Culture: Identity and Communication in Cybersociety,* ed. Steven Jones. Thousand Oaks, CA: Sage Publications.

PART I

Cyber-Social Movements Emerging Online

1

Internet Protests, from Text to Web

Laura J. Gurak and John Logie

Cyberspaces as Protest Sites

From its earliest days, the Internet has been about networking: not just networks of wires and hubs but networks of people. Protests, too, are always about networks, usually networks of people who have a common interest or concern and come together—whether in a physical place, such as in front of a government building, or via a petition or other campaign. No wonder, then, that the Internet has been a useful site for social activism of many forms. But how much do we know about the rhetorical dynamics of Internet protests? Are electronic petitions seen to be just as credible as paper ones? Do mass Web protest campaigns make a difference? Do the speed and reach of online communication bring the same features to electronic protests?

This chapter presents a comparison of two of the earliest Internet-based protests, the cases of Lotus MarketPlace and the Clipper chip, with more recent Web-based protests, such as the protest efforts enabled by sites such as Petitionsite.com and the "Haunting of GeoCities"—a protest against Yahoo!'s attempted appropriation of Web spaces built by "citizens" of GeoCities, the leading purveyor of "free" Web space. In just ten years we have come a long way from the text-based protests of Lotus MarketPlace and the Clipper chip, which relied on newsgroups, e-mail, and the then-novel nature of speed and reach on the Internet. Now, Web pages that go far beyond text in their appeals, using color, sound, images, graphics, and of course, words, still demonstrate the rich opportunities for social action and persuasion in the increasingly visual spaces of the Internet.

Our comparison illustrates that things have certainly changed. But it also points out that some features of online protests, such as speed, reach, problems with fact-checking, new notions of credibility, and traditional power structures, are the same even with the major shift of the Internet from a text-based network to the graphically rich environments found on the World Wide Web. Also, our comparison of "then and now" illustrates that way back when, in the early 1990s, Internet-based petitions and the like were still novel and may have caught people off-guard (such as the CEO of Lotus, who canceled the product after receiving too much e-mail). Today, companies and governments alike take electronic correspondence, including electronic petitions, with a grain of salt. But protests that take advantage of the key features of the Internet, especially the Web's potential for using powerful visual images to reinforce the protest's core message, can still be effective.

The Cases of Lotus MarketPlace and the Clipper Chip

On April 10, 1990, Lotus Development Corporation announced a product called MarketPlace: Households. MarketPlace was to be a direct mail marketing database for Macintosh computers. It would contain name, address, and spending habit information on 120 million individual American consumers from 80 million different households. After MarketPlace was announced, computer-privacy advocates began investigating what they believed was a product that crossed the line in terms of privacy. Although most of the data contained in MarketPlace were already available (the data were provided by Equifax, the number two credit reporting agency in the country), privacy advocates felt that MarketPlace was an inappropriate and invasive application of this data. Having the data so readily available to a mass market of PC users extended what many already felt was a "panopticon" of information sources in the United States, from credit profiles to grocery store checkout scanning systems to government files. Furthermore, the data were provided on the noncorrectable CD-ROM medium. If an entry was in error, it could not be corrected until a subsequent repressing of the database. And although Lotus indicated the privacy protection measures they had put in place, including an encryption scheme so that only "authorized business users" (those who had purchased the program and had somehow been prescreened by Lotus) had access to the data, privacy advocates were not convinced.

From Lotus's first announcement until months after it canceled the product, various electronic bulletin boards and e-mail were full of discussions about MarketPlace. In fact, computer-mediated communication (CMC) was a critical forum in this case. In late November, the *Wall Street Journal* ran a piece about MarketPlace. This story presented Lotus's position as well as the position of Computer Professionals for Social Responsibility (CPSR), an advocacy group that took a position against MarketPlace. Networks were immediately abuzz with discussions of the *Journal* article; soon, debates about the privacy implications of MarketPlace and suggestions for contacting Lotus began to circulate. People posted Lotus's address and phone number, the e-mail address of Lotus's CEO, and information about how to get names removed from the database. Some people posted "form letters" that could be sent to Lotus. Notices were forwarded around the Net, reposted to other newsgroups, and sent off as e-mail messages. In one case, a discussion group was formed specifically to talk about MarketPlace.

One of the most powerful voices within the Lotus protest was that of Larry Seiler, a New England–based computer professional. Shortly after the MarketPlace announcement, Seiler wrote a message that circulated widely via e-mail and Usenet newsgroups:

> Summary: Basically, Lotus is putting together a database, about to be released on CD-ROM in March. It will contain a LOT of personal information about YOU, which anyone in the country can access by just buying the discs. It seems to me (and a lot of other people, too) that this will be a little too much of a big brother, and it seems like a

> good idea to get out while there is still time. Feel free to forward this message to as many people as you know. (qtd. in Gurak 1997, 88–89)

In another letter, Seiler again advanced the notion that simple circulation of the message throughout the Internet might prompt significant enough outrage to build pressure on Lotus. Seiler wrote:

> [P]ass this message along to anyone you think might care. To me, this is not just a matter of privacy. Lotus is going to sell information behind our backs—we are not allowed to dispute their data or even know what it is. Worse, Lotus is going to sell rumors about our income. Still worse, they will do it on a scale never before achieved. This should not be tolerated. Please help to stop Lotus. (qtd. in Gurak 1997, 88–89)

While Seiler's messages do not outline specific strategies for protest, they were nevertheless resoundingly effective, triggering waves of ad hoc action by sympathetic Netizens.

During the following Internet-based protest, over thirty thousand people contacted Lotus and asked that their names be removed from the database. The product, which had been scheduled for release during the third quarter of 1990, was, ultimately, never released. On January 23, 1991, Lotus issued a press release announcing that it would cancel MarketPlace: Households because of "public concerns and misunderstandings of the product, and the substantial, unexpected costs required to fully address consumer privacy issues" (Gurak 1997, 19). In the end, many acknowledged the role of Internet-based networks in stopping the release of MarketPlace. Some subsequently called it "[a] victory for computer populism" (Winner 1991, 66).

Four years later, a similar online action took place. The Computer Security Act of 1987 required that the National Institute for Standards and Technology (NIST), a federal standards-setting organization under the Commerce Department, develop a new national standard for computer encryption. This standard would replace the existing data encryption standard, known as DES, in response to the need for a more sophisticated approach. Unlike the proposed Clipper standard, which requires two keys (each held by a different agency), the DES involved a single key to both encrypt and decrypt a message; by 1987, this design was considered outdated and not sophisticated enough to support the continuing "information revolution."

NIST thus followed the directive of the 1987 Computer Security Act and began work on a new federal encryption standard. To do so, they turned to the National Security Agency (NSA), described as "the United States' most secretive intelligence organization" (Markoff 1993, D1). The NSA proceeded to develop an escrowed encryption standard (EES), which would be implemented in a chip that came to be known as Clipper. This chip could be inserted into a telephone handset or fax machine. On April 16, 1993, President Clinton proposed the EES as the new encryption standard.

This announcement triggered immediate concern among privacy advocates. The lack of concern on the government's part for public input caused groups like CPSR and the Electronic Frontier Foundation (EFF) to begin sounding alarm bells. Media coverage began appearing, highlighting the ideological split between privacy advocates and the government's proposal. In addition, computer and telecommunications industries, aware of the growing markets for communication technology, were troubled by the implications of the Clipper chip. No foreign companies, they argued, would want to purchase products using encryption schemes that could be unscrambled by U.S. investigative agencies. The Clipper standard would be a severe blow to U.S. exports; *Forbes* magazine called it "really a dumb idea," suggesting that "[h]igh-tech exports will be devastated" (Forbes 1994, 26). Industry representatives thus joined with privacy advocates to voice continuing opposition to the Clipper chip.

As in the Lotus MarketPlace case, cyberspace was an important forum for discussions, debates, and protests over the Clipper chip. Information moved across the Internet via e-mail, Usenet newsgroups, and discussion lists. Special ftp sites were set up to house important Clipper-related documents, and before long the Internet was "buzzing with talk of insurrection" about Clipper (Markoff 1994a, D4). Some of this buzz took the form of hot-tempered eruptions. One posting to the *Computer Privacy Digest* read: "DEFEAT THE BIG BROTHER PRO-

POSAL! JUST SAY F!CK NO TO THE PRIVACY CLIPPER!" (qtd. in Gurak 1997, 63). But the more popular forms of protest clearly took into account the U.S. government as the ultimate audience, and for this reason, sample form letters and electronic petitions became increasingly dominant as protest efforts progressed. The most popular of these petitions could be "signed" by simply typing in one's information and sending the file back to CPSR. This petition, while distributed via e-mail, took on the form and conventions of a paper-based letter, complete with an address block reading "The President/The White House/Washington DC 20500" and reflected the more careful prose of a reflective, collaborative composing effort, as opposed to Seiler's relatively raw, emotionally charged language. This petition garnered anywhere from forty thousand to fifty thousand signatures.[1] Despite these industry and advocacy efforts, the Clinton administration officially adopted the Clipper as a federal information-processing standard for voice communications on February 4, 1994.

What We Learned from These Text-Based Cyber-Actions

From a rhetorical perspective and from the perspective of Internet studies, these two cases have much to teach us about online actions. First, we learned that on the Internet, exigencies come together quickly and can snowball in a matter of days or even hours. In today's webbed world of online news, hoaxes and humor, and the like, this feature may seem almost commonplace. But it is only recently that we have had such a technology as to allow a social effort or action to form, with tens of thousands of participants, in such a short time. In both the MarketPlace and Clipper cases, discussions and protests got off to a quick start, within twenty-four hours of the announcement of each technology.

By the end of the 1980s, discussions of computers and personal privacy were in the public eye. Lotus MarketPlace acted as a catalyst around which the exigence could then focus as individuals began using the Internet to talk about MarketPlace. MarketPlace and, later, Clipper thus gave people the "mobilization exigency" around which to organize these concerns about computer privacy; such exigence has been argued to be a feature that can help distinguish "the rhetorical situation of movements" and protests from other rhetorical situations (Smith and Windes 1976). As one privacy advocate put it, the MarketPlace protest community was "like kindling waiting for a spark" (Rotenberg 1992); both MarketPlace and later Clipper provided needed sparks.

Another lesson from these two early cases has to do with the power and potential of online communities. Participants were able to assume that others in the newsgroups or lists understood certain technical concepts and agreed with certain premises. The highly specialized virtual spaces on the Internet make it easy to join a community and quickly understand and assume this community *ethos*; a newsgroup focused on computer privacy, for example, is most likely to be inhabited by participants who are concerned about privacy and want to protect their rights. Often, participants often do not have to spend time making introductory remarks or defending the premises of their statements. This "instant *ethos*" makes it easy to reach many individuals of similar values in short order and, when combined with online delivery, allowed for both protests to focus quickly. Assumptions about technical knowledge and computer privacy in both cases allowed for the creation of short, direct messages that assumed the community *ethos* and would appeal to the readers of these messages. In addition, an authoritative and ironic voice offered a strong challenge to Lotus's or the government's claims and invited other readers to join the debate.

Both cases also illustrate the way in which the Internet's nonhierarchical structure allows individuals to bypass "standard procedure" and reach out to each other. As the debates continued beyond their initial stages, certain texts became widely reposted and distributed. In the protest over MarketPlace, the most prominent posting was "the Seiler letter," which, although initially posted to only a few sites, was soon widely available on the Internet as participants copied and reposted it. In the Clipper case, CPSR's electronic petition and letter to stop Clipper were also widely distributed. In both cases, these bottom-up texts became representative of the debate at large and created cohesion among participants across the Internet.[2] Once participants learned of MarketPlace and later Clipper, they could and did easily use e-mail to write directly to the CEO of Lotus or the President of the United States, bypassing traditional hierarchical structures. In addition, some participants utilized the ability in cyberspace to write anonymous postings, such as a purported internal press release from Lotus and, in the Clipper case, purportedly confidential information from the manufacturer of the chips. These anonymous postings also circumvented traditional gate-keeping structures and allowed information to circulate widely, under the radar so to speak, creating what one participant called "an electronic wave going around the world."

The Lotus and Clipper cases also presented early warning about when and how one can judge the credibility of information from the Internet. Much of the material about Lotus was angry and critical of MarketPlace, and it contained inaccurate and often hyperbolic information about the MarketPlace product; many of these inaccuracies came about as a result of the bottom-up method of posting and reposting. The electronic petition against Clipper, on the other hand, exhibited a highly professional *ethos*, one made all the more credible by CPSR's name, which was prominently attached to the petition. In general, information remained relatively accurate throughout the Clipper debate, in large part because most of this information was circulated from the top down through organizations such as the EFF and CPSR.

Web-Based Protests: Petitionsite.com

The kinds of text-based actions that took place informally in the Lotus Marketplace protest and in a more structured way within the Clipper protests are now being made available to much larger audiences, and in a considerably streamlined fashion, via sites like Petitionsite.com. This centralized "petition clearinghouse" lists dozens of online petitions sorted by category (e.g., "Health," "Animals/Environment," "Politics and Government–State"), by sponsoring organization, and by urgency. Virtually all of the steps that users developed through trial and error via e-mail and within newsgroups have been wholly automated.

Visitors to specific petitions housed at Petitionsite.com are presented with a brief overview of the topic of the petition, with a direct link to the full text. The most prominent element of the pages is a graphic encouraging site visitors to sign petitions even before they have likely read the full text of the petition (see Fig. 1.1). Visitors also encounter a table listing the twenty-five most recent signatories to a petition, most

Figure 1.1 · Petitionsite.com's "signature block."

listing a short rationale for their participation. While e-mail–based protests often contain long skeins of e-mail addresses, documenting a petition's movement throughout the Internet, these e-mail addresses are often cryptic and functionally anonymous. Petitionsite.com offers visitors relative specificity, reinventing one of the elements of paper-based petitions for online spaces. Unless signatories choose to remain anonymous, their names are attached to specific comments, helping these petitions to serve as community-centered documents, in much the same way that paper-based petitions might grow out of collective discussion.

Petitionsite.com also offers a Web-based mechanism for routing petitions to friends, colleagues, or, for that matter, anyone who has an e-mail address. While Petitionsite.com offers a substantially automated approach to Internet protests, it is still grounded in print paradigms. As with the Clipper chip protest, Petitionsite.com produces more efficient versions of print petitions, thereby expanding the potential pool of participants to the limits of the digital divide.

From Web-Based to Web-Enabled: The Yahoo!/GeoCities Protest

In 1994 the Internet-based start-up GeoCities began offering one megabyte of computer storage space to anyone willing to build and maintain a Web site within any of dozens of GeoCities "neighborhoods." GeoCities described those who developed Web spaces in these neighborhoods as "homesteaders," and the company's business model depended upon people building on GeoCities's property, and thereby enriching its value. GeoCities did not initially require that users cede space in or around their Web pages for advertisements, instead envisioning a model similar to the model billboard companies have followed for years. GeoCities counted on homesteaders first to create attractive sites, then attractive neighborhoods, and GeoCities simply maintained the right to sprinkle the thoroughfares with advertisements.

Over time, GeoCities adjusted this balance, first requiring homesteaders to incorporate a link to the GeoCities homepage (although this "requirement" was loosely enforced and widely ignored) and then attaching that bane of the Internet, the pop-up advertisement, to each homestead. Each of these developments diminished the attractiveness of homesteading at GeoCities—and a few homesteaders left as the ads grew more intrusive and aggressive. But the number of new homesteaders moving in more than counterbalanced the disgruntled few who departed, and GeoCities maintained its status as one of the most popular purveyors of "free" Web space throughout these changes to the original bargain. According to 1999 press reports, the whole of GeoCities, with, by then, thousands of neighborhoods encompassing hundreds of thousands of Web sites, constituted the fifth most popular destination on the World Wide Web.

So popular was GeoCities that it's not GeoCities anymore. In January 1999, the Internet giant Yahoo!, flush with capital from its initial public offering, bought GeoCities in a stock swap valued at somewhere between $3.5 billion and $5.2 billion. Shortly thereafter, GeoCities became "Yahoo! GeoCities" and the initial GeoCities "Homestead Act" received a dramatic revision from the new corporate parent.

On June 25, 1999, Yahoo! posted an amendment to the GeoCities terms of service, which read, in part:

> 8. CONTENT SUBMITTED TO YAHOO!
> By submitting Content to any Yahoo property, you automatically grant, or warrant that the owner of such Content has expressly granted, Yahoo the royalty-free, perpetual, irrevocable, non-exclusive and fully sublicensable right and license to use, reproduce, modify, adapt, publish, translate, create derivative works from, distribute, perform and display such Content (in whole or part) worldwide and/or

> to incorporate it in other works in any form, media, or technology now known or later developed. (Townsend 1999d)

This passage, bearing the unmistakable stamp of "corporate counsel," effectively claimed the GeoCities homesteads as Yahoo!'s intellectual property. The reference to forms, media, or technology "now known or later developed" borrows its language from Section 102 of Title 17 of the United States Code, the section defining copyright. Any legally aware reader would reasonably conclude that with this passage, Yahoo! was claiming ownership of copyright for all of the material homesteaders had posted to Yahoo!'s servers.

In the original GeoCities Terms of Service, GeoCities reserved the right to prohibit patently offensive pages and commercial pages, but it did so with language that made it clear that GeoCities was not claiming ownership over content. The revision constituted a dramatic expansion of GeoCities's claim to the "homesteads." For most Web designers, the term "homepage" is a fairly inert term, designating nothing more than a page's centrality within a larger network of pages. But for the GeoCities homesteaders, "Personal Home Page" had come to signify much more than that.

When Yahoo! revised the GeoCities Terms of Service, it not only posted the revision to the GeoCities Web site, but also force-fed the revision to anyone attempting to revise their GeoCities-hosted Web pages and required an electronic "signature" agreeing to the new Terms of Service before any changes to the site could be made. This approach prompted outrage among the homesteaders, and their response to Yahoo!'s actions was, ironically, indicative of GeoCities's success in establishing real communities on the virtual frontier. Within a matter of hours, hundreds of members of this community called out to one another, united, and set about developing a strategy for persuading Yahoo! GeoCities to revert to something like the original GeoCities bargain.

One of the most powerful of the homesteaders' early calls for action was made on June 30, 1999, by a self-described "contract internet [*sic*] software developer that happen[ed] to be between contracts" named Jim Townsend. Townsend established a Web site called "come to/boycottyahoo," which quickly became the central information conduit for homesteaders wishing to protest Yahoo!'s actions. In the first of a series of daily editorials posted to this site, Townsend outlined a series of possible strategies for homesteaders' revolt:

> Stop using Yahoo. Boycott them, and all of their properties. This includes Yahoo.com, GeoCities.com and Broadcast.com. Don't buy products from merchants at shopping.Yahoo.com and let them know why!
>
> Let Yahoo know that you won't tolerate this! Email copyright @yahoo-inc.com and demand that they let you remove your content AND demand that they remove it from their databases AND demand that they won't use it without your express consent.
>
> Move your site to one of GeoCities' competitors (see list at left).
>
> Link to this page from your site (your new site, if you're moving from GeoCities) and let your visitors know that this issue is important to you as a webmaster, even if you don't use GeoCities yourself!
>
> If you're a user and you use a site which is located at GeoCities, let the owner of that site know that he or she has just signed over his or her entire work to these goons and they should leave the service at once!
>
> If you are a graphics professional, design logos which don't infringe on Yahoo's copyrights but will give viewers the message that they are under boycott (think of the ribbon campaign to defeat the CDA [Communications Decency Act] last year).
>
> If you are a reporter, or know one, or know how to get in touch with one, grab their ear and let them know this is going on.
>
> Most of all, get up and TAKE ACTION. Yahoo and their "properties" don't make any money if you don't use their services, so don't! Tell your friends and family, point them here so they can inform themselves! (Townsend 1999d)

As in the Lotus and Clipper cases, people *did* take action, and quickly. Taking up Townsend's challenge, graphically talented GeoCities members were soon implementing a sophisticated form of protest that took advantage of the Web as a malleable composing space (see Fig. 1.2). Within hours, Townsend's site was offering dozens of banner ads and

Figure 1.2 · Exemplary parodic "banner ads" generated during the Yahoo!/GeoCities protest.

graphics composed by disgruntled homesteaders for free use by others participating in the protest. Many of these ads incorporated motion and slyly parodied GeoCities's own typographic "house style." Others invented parodic slogans such as "Yahoo and GeoCities: We own everything so you don't have to!"

Across GeoCities, Web pages filled with downloadable protest graphics sprouted like proverbial mushrooms. Townsend was careful to sensitize those submitting protest graphics to the attendant intellectual property issues, writing: "Make sure its [sic] entirely your own creation, and make sure you want other Web sites to use it" (Townsend 1999a). By so doing, Townsend was encouraging others to freely distribute specific pieces of intellectual property in order make the larger argument that Web composers ought to be able to both own and control the distribution of their work. By selectively giving their work away, these designers were implicitly arguing that they, *and not Yahoo!*, were the ultimate owners of their homesteads' contents. While this strategy has a certain elegance, it was soon complemented by an even more elegant protest technique—the "Haunting" of GeoCities.

During the Haunting, participating homesteaders revised their homepages to feature a dark "battleship gray" background that often obscured or concealed their pages' original content (see Fig. 1.3). Some homesteaders took the additional step of withdrawing their content altogether and replacing it with content decrying Yahoo!'s actions. Typically, these pages featured the date that the homesteader joined the boycott, links to other boycott sites, and a catalog of that homesteader's particular complaints about Yahoo!. While these complaints are, predictably, somewhat varied, the sites were usually careful to specify Section 8 as the target of the protest.

The speed with which the homesteaders protested the revised Terms of Service was remarkable. Yahoo! posted the revision to its Web site on June 25. On June 29, Townsend received a letter from a friend asking him about the revised Terms of Service agreement. He posted his rallying cry to the newly built "cometo/boycottyahoo" Web site the *following day*. From that point forward, Townsend's site became the central distribution point for boycott graphics, and the central link to the Haunting pages. The popularity of Townsend's site is attested to by the high number of Haunted sites that linked to Townsend's site and identified it as the home of the boycott. By 4:30 P.M. on June 30 *WIRED* magazine's online edition had published one of the first news stories on the protest, citing Townsend's Web site as one of its main sources. Also that afternoon, Yahoo! was already backing away from the revised Terms of Service, posting a further revision. Townsend posted a critique of the new revision to his Web site within a matter of hours.

By July 1, the *New York Times* print edition published a brief article that parroted the *WIRED* news story and quoted Townsend's two-day-old Web site. Buoyed by the publicity, Townsend began calling for a "Homesteader's Bill of Rights" and issuing Web-based press releases that called for Yahoo! to eliminate the remaining problems with the Terms of Service revisions by Independence Day. At this point, Townsend's site had received over three hundred thousand hits, repre-

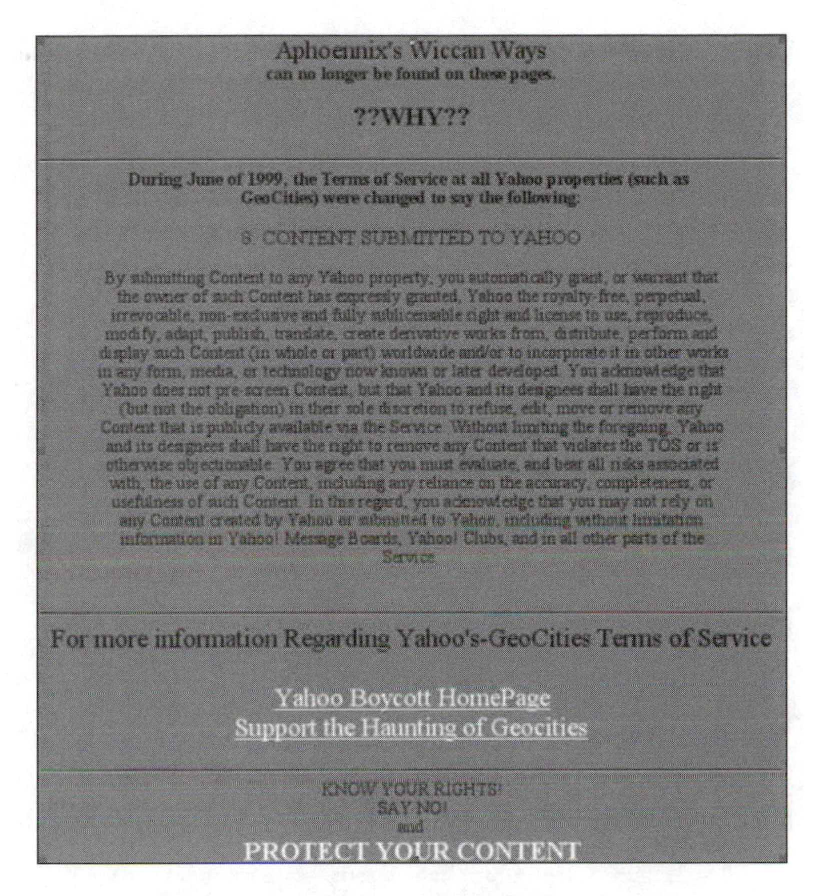

Aphoennix's Wiccan Ways
can no longer be found on these pages.

??WHY??

During June of 1999, the Terms of Service at all Yahoo properties (such as GeoCities) were changed to say the following:

8. CONTENT SUBMITTED TO YAHOO

By submitting Content to any Yahoo property, you automatically grant, or warrant that the owner of such Content has expressly granted, Yahoo the royalty-free, perpetual, irrevocable, non-exclusive and fully sublicensable right and license to use, reproduce, modify, adapt, publish, translate, create derivative works from, distribute, perform and display such Content (in whole or part) worldwide and/or to incorporate it in other works in any form, media, or technology now known or later developed. You acknowledge that Yahoo does not pre-screen Content, but that Yahoo and its designees shall have the right (but not the obligation) in their sole discretion to refuse, edit, move or remove any Content that is publicly available via the Service. Without limiting the foregoing, Yahoo and its designees shall have the right to remove any Content that violates the TOS or is otherwise objectionable. You agree that you must evaluate, and bear all risks associated with, the use of any Content, including any reliance on the accuracy, completeness, or usefulness of such Content. In this regard, you acknowledge that you may not rely on any Content created by Yahoo or submitted to Yahoo, including without limitation information in Yahoo! Message Boards, Yahoo! Clubs, and in all other parts of the Service.

For more information Regarding Yahoo's-GeoCities Terms of Service

Yahoo Boycott HomePage
Support the Haunting of Geocities

KNOW YOUR RIGHTS!
SAY NO!
and
PROTECT YOUR CONTENT

Figure 1.3 · A typical Haunted Web site during the Yahoo!/GeoCities protest.

senting over 15 percent of GeoCities's estimated 4.6 million users. Townsend's site also listed over five hundred Web pages that had removed their initial content and replaced it with Haunted content. The creativity exhibited in the numerous parodic banners found on Townsend's site clearly prompted similarly creative efforts by those Haunting their sites (see Fig. 1.4).

By July 2, Townsend's pages had received more than a million hits, and many of his visitors altered their GeoCities sites to participate in the Haunting. Additional networks for the production and distribution of protest graphics and techniques began branching from

Figure 1.4 · A graphically intensive Haunted site.

Townsend's sites, including a central site that explained how and why to construct a Haunted GeoCities page. Townsend cannily guided the protest, encouraging visitors to monitor media coverage of the protest and forward links to his site, so that this coverage could be incorporated into the ever-expanding corpus of protest-related discourse. On each passing day visitors to Townsend's site were greeted by, at minimum, a freshly written editorial (Townsend described them as "screeds") and revised links to press releases, protest pages, Haunted sites, and media coverage. By July 5, the protest story had been picked up by Reuters, which meant it was appearing on Yahoo!'s own news site, which depends heavily on the Reuters newsfeed. Townsend, by this time the generally recognized leader of the protest, was urging an increasingly large, increasingly angry group of protesters to continue fighting the inadequate revisions with creative, challenging discourse, *not* the electronic equivalents of pranksterism and terrorism.

> Keep your chins up and your eyes and ears open for ways to increase the pressure against Yahoo (legally, folks, lets not start spamming or any of that garbage; we have the high moral ground here lets not throw that away). (Townsend 1999b)

Three days into a snowballing protest effort, Townsend was articulating the importance of maintaining the protestor's *ethos*. Townsend was thereby promoting a reinvention of rhetoric tailored to virtual and electronic spaces.

Townsend and his fellow protesters were, by this time, regularly being interviewed by television stations, and virtually every major online news outlet had run a story on the protest. Yahoo!'s competitors were aggressively capitalizing on the homesteader's unhappiness, advertising their own nonrestrictive terms of service agreements to users considering relocating from Yahoo! GeoCities. In short, Yahoo! GeoCities was in the midst of a public relations nightmare, and Townsend's able management of the protest seemed certain to ensure that things would only get worse for Yahoo! GeoCities until something like the original bargain was restored. Yahoo! was now faced with the prospect of watching its neighborhoods become blighted as some homesteaders deserted and others darkened their fractions of the New Frontier. Faced with the prospect of losing the brand equity in the GeoCities name, Yahoo! did the sensible thing, and on July 6, 1999, the corporation published a third revision to the Terms of Service. This radical revision represented a shift in both tone and content from the confusing legalese that had characterized the previous revisions:

> 7. CONTENT SUBMITTED TO YAHOO! GEOCITIES
> Yahoo does not claim ownership of the Content you place on your Yahoo GeoCities Site. By submitting Content to Yahoo for inclusion on your Yahoo GeoCities Site, you grant Yahoo the world-wide, royalty-free, and non-exclusive license to reproduce, modify, adapt and

> publish the Content solely for the purpose of displaying, distributing and promoting your Yahoo GeoCities Site on Yahoo's Internet properties. (Yahoo! GeoCities 1999)

By 9 P.M., PST on July 6, Townsend had issued a celebratory editorial declaring the boycott/Haunting over:

> Tonight, July 6th, 1999, Yahoo published a revised Terms of Service agreement specifically for its GeoCities property and its homesteaders.
>
> The new GeoCities ToS addresses each of these concerns in a clear, positive and concise manner which may well serve as a blueprint for similar Terms of Service agreements throughout the young internet community. (Townsend 1999c)

Yahoo! began requiring users to sign on to the initial revision of the GeoCities Terms of Service on June 28, 1999. The boycott, the Haunting, and the general program of collective protest actions prompted a reversal of Yahoo!'s decision in nine days. While Jim Townsend's Web site anchored the boycott, the most powerful and persuasive action taken during this nine-day period was the Haunting, in which, one by one, virtual homesteads, and then virtual neighborhoods, turned the lights out. GeoCities had built the roads, but all of a sudden nobody was home. By any measure, this was a remarkably effective protest action.

What We Are Learning from Web Protests

While the Yahoo!/GeoCities protest features the same kind of furiously snowballing exigency we first encountered in the Lotus and Clipper cases, the Web is also capacious enough to accommodate protests with more elongated timelines. Indeed, a significant percentage of the petitions found on Petitionsite.com are designed to call attention to little-known issues, by trading on the increasing popularity of Petitionsite itself. Thus, while text-based protests and the Haunting depended on grassroots networking efforts, Web protests need not necessarily function in the same way. That having been said, without supporting information-distribution efforts, these sites surrender the breathtaking speed that has, to date, characterized the most successful Internet protests.

Web protests reinforce our sense that real communities can and do take root in Internet-based spaces. While the "residents" of GeoCities's thousands of neighborhoods were probably more diverse and disparate than the residents of typical terrestrial neighborhoods, they nevertheless united in defense of their rights to their virtual properties. Whole neighborhoods within GeoCities altered their appearance, "darkening their doors" to protest the perceived injustice of a shift in their governance. It is hard to imagine a terrestrial neighborhood presenting a more united front. Paradoxically, GeoCities ultimately achieved its initial goal of promoting the formation of real communities in cyberspace by threatening the residents of those communities. In so doing, it prompted a wave of interaction, interlinking, and exchange among the members of its neighborhoods that otherwise might never have occurred. To a lesser extent, the individuals adding their signatures and rationales to the petitions at Petitionsite.com are building more limited connections with one another, forming loose communities centered around issues rather than "residency."

Relative to text-based protests, Web protests are relatively anonymous, and this potentially raises serious credibility concerns. Many of the Yahoo!/GeoCities protest graphics were stripped of attribution as they circulated among the protesters. Further, because many of the protest "banner ads" closely aped existing Yahoo! or GeoCities banner ad campaigns, there was the remote potential for these graphics to be mistaken, at least at first blush, as communication from the corporation. The Haunted Web pages, by comparison, offer a reasonably straightforward establishment of ethos, with protesters often directly citing the offending Terms of Service agreement and voicing their personal objection to these terms. The other key means of establishing ethos within Web spaces is the hyperlink, and the Yahoo! GeoCities protesters used this Web feature to great effect, turning each Haunted page into a node in an enormous network, routing visitors to Townsend's central distribution point and to others who were contributing to or expanding the protest's scope. By contrast, Petitionsite.com changes the dynamics of the petition as a genre because it allows for anonymous "signatures." In a print context, it is never entirely possible to anonymously sign a petition. Whether the addition of anonymous "signatures" to these Web-based petitions results in a closer approximation of a given community's true views on a subject remains to be determined. But it is already clear that because Petitionsite.com often invokes print paradigms, its petitions are perceived as less credible than print-based petitions, which are not subject to anonymous spoofing, hacking, and co-optation on the scale enabled by electronic media.

Finally, these Web protests both illustrate the degree to which the Web intensifies the Internet's nonhierarchical structure, which was also a striking feature of the text-based protests. Both Petitionsite.com and the Yahoo! GeoCities protest pages feature mechanisms for the rapid circulation of protest materials, but these mechanisms often transmit messages without also transmitting the ultimate source of those messages. On both sites, protest participants seem willing to defer questions of credit and attribution in exchange for the establishment of a collective, community-centered ethos. Within these sites, leaders tend to be retroactively identified. While protests are active, even those in positions of leadership tend to identify themselves as "participants." This is especially true on Petitionsite.com; it is often difficult to locate the author of a particular petition. But while Web protests substitute networks of exchange for traditional organizational structures, they are potentially more efficient than text and e-mail-based protests. While members of privacy-oriented newsgroups were inundated with repeat postings of the same material during the Lotus Marketplace and Clipper chip protests, the Web serves as a two-way check on such redundancy. The presence of a particular Web site obviates the need for repeat posting of information, and movement within a complex network of protest sites is circumscribed by the patience of individual visitors. On the other hand, sites like Petitionsite.com arguably lower the bar for online protest, enabling cranks who might not otherwise be able to circulate their messages to take advantage of a streamlined distribution mechanism and to associate their idiosyncratic protests with more legitimate efforts by establishing misleading or specious links.

Our review suggests that most online protests far outstrip their print counterparts in terms of speed and reach. Internet protest efforts are often measured in days and hours, whereas paper-based protest efforts move no faster than postal carriers or community activists. But this speed is achieved at the expense of the kinds of verification and vetting that have often been applied in paper-based contexts. And while Internet protests often boast participation numbering in the tens of thousands, the level of interaction underpinning this participation can and should be questioned. Indeed, the nonhierarchical nature of the Internet often makes it difficult to establish what could or should constitute meaningful participation in a protest action.

But while Web-based protests are, at least for the time, superficially less credible than their print and e-mail-based predecessors, it is also clear that Web-centered protests routinely produce sophisticated arguments that take full advantage of the rich hypertextual and visual opportunities offered within the World Wide Web. As such, these arguments represent rich examples of rhetoric in action, and of the ability of motivated people to craft messages uniquely suited to the spaces and communities they value.

Notes

1. Reports vary on the actual number of signatures to the petition. For example, EPIC reported forty-seven thousand signatures, while the major print media reported anywhere from forty-five thousand to fifty thousand. Depending on how many versions of the petition were circulating and how many versions were ultimately delivered to the White House, each of these figures could in a sense be true. Even after a single version was delivered, other versions might still be circulating and gathering signatures. This variation in signature count points out a significant difference between traditional paper texts and interactive electronic texts; the fixed nature of the former is replaced by a dynamic quality in the latter.

2. Although, as the analysis reveals, each case exhibited a subtle yet important difference in this regard: MarketPlace was very much a bottom-up action, driven by a few individual postings such as the Seiler letter, which held such strong appeal that they were reposted widely and thus provided a cohesion based on community consensus. The Clipper debate was a blend of both top-down *and* bottom-up structures, organized and continually maintained by the postings of CPSR and EFF but also maintained via subsequent repostings as well as postings from many individuals.

References

Forbes, Malcolm S., Jr. 1994, April 14. "High-Tech Snoops." *Forbes,* 26.

Gurak, Laura J. 1997. *Persuasion and Privacy in Cyberspace: The Online Protests over Lotus MarketPlace and the Clipper Chip*. New Haven, CT: Yale University Press.

Markoff, John. 1993. May 6. "Big Brother and the Computer Age." *Business Day,* D1.

——. 1994a, Feb. 13. Cyberspace under Lock and Key. *New York Times*, D4.

Rotenberg, Marc. 1992. Personal interview. December 4, 1992.

Smith, Ralph R., and Russel R. Windes. 1976, Fall. "The Rhetoric of Mobilization: Implications for the Study of Movements." *Southern Speech Communication Journal* 42:1–19.

Townsend, Jim. 1999a. "Boycott Graphics." Online. http://www.globalterror.com/messagezone/boycottyahoo/boycottimages.htm (May 23, 2000). [No longer online.]

——. 1999b. "Boycott Yahoo Editorial–Day 7." Online. http://www.globalterror.com/messagezone/boycottyahoo/editorial-day7.htm (May 23, 1999).

——. 1999c. "It's Over!!!" Online. http://www.globalterror.com/messagezone/boycottyahoo/boycottyahoo.htm (May 23, 2000). [No longer online.]

——. 1999d. "Yahoo and GeoCities Have Gone Too Far." Online. http://www.globalterror.com/messagezone/boycottyahoo/original.htm (May 23, 2000). [No longer online.]

Winner, Langdon. 1991, May/June. "A Victory for Computer Populism." *Technology Review*: 66.

Yahoo! GeoCities. 1999. "Terms of Service." http://docs.yahoo.com/info/terms/geoterms.html (May 23, 2000).

2

Indymedia.org

A New Communications Commons

Dorothy Kidd

On September 11, 2001, I first heard about the attacks on the World Trade Center and Pentagon from conversation with the early morning regulars at my local café in San Francisco, California. Returning home, I quickly turned to three other sources of information—network television, KPFA-FM radio, and the World Wide Web's indymedia.org, the site of the Independent Media Center (IMC).

I flipped around the TV and saw that all the major networks had canceled their regular advertising-driven programming to provide round-the-clock coverage of the attacks and their aftermath. For the following month, the networks provided more hard news about government, military, national, and especially international affairs than had been seen in decades. However, as the Project for Excellence in Journalism Report (CJR) confirms, the shift in news agenda was only in the subject in focus, and not the overall approach (2001, 1). The selection and emphasis of news content were still within the very tight framework that favors U.S. political and economic elites (Ryan, Carragee, and Meinhofer 2001; Smith et al. 2001). There was little criticism of the policies and actions of the U.S. military and the Bush administration; little discussion of alternative political views, especially of peace movements; and little talk of the international political and economic context that might help explain this crisis (Solomon 2001).

My second source, Pacifica Radio, began immediately to explore that larger context, absent in the corporate media. The morning of September 11, Berkeley's KPFA-FM, interrupted their regular programming to run a live feed from New York's new program, "Democracy Now," broadcasting only blocks away from Ground Zero. The "special" team provided coverage of the impact on New York and Washington, as well as background interviews about the history of Afghanistan, and especially of U.S. relations in Afghanistan and the Middle East. For the next month, "Democracy Now" and KPFA continued the dual focus, covering stories from perspectives missing in network TV coverage: the peace protests in New York, San Francisco, and around the world, and the voices of victims' families who stood for peace; illegal immigrants lost in the Twin Towers; New Yorkers trying to reconstruct their lives amidst the environmental and fiscal devastation; individuals from the Arab and South Asian communities who were targets of discrimination, violence, and police detention in the United States; as well as journalists, activists, and scholars from around the world.

One might imagine that Pacifica would step up to provide this coverage. Started by pacifists after the Second World War, the first station, KPFA, was founded to counter the build-up of the U.S. military-industrial complex and to challenge the monopoly control and commercialization of the broadcast media. Since then, KPFA and the noncommercial listener-supported Pacifica Network it formed have modeled a communications resource that draws from "sources of news not commonly brought together" and on-air dialogue between people of widely differing political and philosophical views (Land 1999; Lasar 1999). Pacifica had been a leading independent media voice for several decades, covering the McCarthy era, the Vietnam War, and the civil rights movement, as well as the rise of the new social movements of African Americans, Native Americans, women, Latino Americans, lesbians, and gays.

However, in the 1990s, a series of disputes over the makeup of programming and leadership had precipitated a major crisis. Supporters and staff were embroiled in lawsuits, workplace disputes, and public actions with the national management. By the fall of 2001, a series of firings, bans, and strikes led to a decline in news and other programming. Amy Goodman, host of "Democracy Now," was banned from the New York Pacifica station, WBAI, and she and her staff were working as independent contractors, sharing a studio with other media activists in a local community television center. While "Democracy Now" still provided cutting-edge commentary and dissent, a new generation of social movements, organized around anti-corporate globalization, had given rise to a new critical medium of independent news and commentary.

"Don't Hate the Media—Become the Media"[1]

My third choice was to browse IMC's Web site, which I have followed through meetings, conferences, collaborations, and interviews since its beginnings. Within the first few days following September 11, the site featured street-level descriptions of peace vigils and demonstrations in the United States and internationally. On the Israeli site, I also found a strong comment from a human rights activist, condemning the attack and countering the corporate media's attempt to link it to Palestinians. All served as important correctives to the barrage of support on TV for the U.S. government's military build-up.

Indymedia is made up of over sixty autonomously operated and linked Web sites in North America and Europe, with a smaller number in Africa, Latin America, and Asia. The first IMC was started in Seattle in 1999, just before the encounter between the World Trade Organization (WTO) and the social movements opposed to its policies. Early on in the counter-WTO planning, several different groups had recognized the strategic importance of making an "end-run around the information gatekeepers" to produce their own autonomous media (Tarleton 2000, 53). They were well aware of the limitations of depending on the corporate media to provide coverage, especially the necessary analyses and context for the complex changes threatened by the WTO regime. In fact, before the event, only a handful of articles in the U.S. corporate media had discussed the implications of the WTO meetings.[2]

The IMC would not have been possible without the convergence of new levels of social movement organization and technology. In three short months in the fall of 1999, and with only $30,000 in donations and borrowed equipment, Seattle organizers created a "multimedia peoples' newsroom," with a physical presence in a renovated downtown storefront and in cyberspace on the Web (Tarleton 2000, 53). The IMC enabled independent journalists and media producers of print, radio, video, and photos from around the world to produce and distribute stories from the perspectives of the growing anti-corporate globalization movement. The IMC was the child of a collaboration between local housing and media activists; journalists, independent media producers, and media and democracy activists from national and international arenas; and local, national, and international organizations active in the burgeoning anti-corporate globalization movement.

Second, the Seattle IMC drew from the technical expertise and resources of computer programmers, many of whom came from the open-source movement. While Bill Gates of Microsoft played a major role in bringing the WTO to Seattle, Rob Glaser, who made his millions at Microsoft, donated technical support and expertise, and in particular the latest streaming technologies, to the indymedia Web site. "From the standpoint of all these independent media, the WTO couldn't have picked a worse place to hold their meeting," according to local media activist Bob Siegel. "I mean it's Seattle—we've got all the techies you'll ever want. . . . It's perfect that the WTO came here. Perfect" (quoted in Paton 1999, 3). Indymedia.org allowed real-time distribution of video, audio, text, and photos, with the potential for real interactivity through "open publishing," in which anyone with access to the Internet could both receive and send information.

In just two years, the IMC network has become a critical resource for activists and audiences around the world, providing an extraordinary bounty of news reports and commentaries, first-person narratives, longer analyses, links to activist resources, and interactive discussion opportunities from around the world. In the beginning, they focused primarily on the anti-globalization mobilizations at the multilateral summits of neoliberal governance. At each of these meetings, they provided innovative international coverage, which often included collaborative initiatives with other media and social-movement activists. In the last year, and particularly since September 11, the network has added several new member sites and widened the scope of its coverage to include local, national, and international campaigns concerning anti-corporate globalization.

In this chapter, I demonstrate how the independent media centers are a new watershed in a historical continuum of radical activist media, in which media activists have continually created new communications resources and challenged the enclosure of the communications commons.[3]

> The Seattle IMC and the growing Independent Media Center Network represent a new and powerful emerging model that counters the trend toward the privatization of all public spaces by expanding our capacity to reclaim public airwaves and resources. (IMC Brochure 2001, 1)

The IMC constitutes a new commons regime, relatively autonomous from the direction of the corporate and state media, in which unpaid workers share cyber and real territories, labor time, and communications technologies, techniques, and techne.

Lessons from the Commons and Enclosures

The concepts of the commons and enclosures date back to a conflict in England five hundred years ago. However, the terms have recently been given new currency in the debate over globalization and development and global communication. There are broadly three sets of meanings in use, which correspond with the three sets of social actors involved in these debates—the capitalist market, the state, and social movements (Sénécal 1991). Neoliberal economists invoke the "tragedy of the commons," arguing, as did the first feudal landlords, that the resource should be enclosed under corporate control in order to stop its unregulated overuse and make it more efficient and that nonconforming practices should be criminalized (Travis 2000).[4] The second school describes the commons as a "public" resource, which should be managed by state or multilateral international institutions, or public-private partnership.

My own perspective derives from groups within the social movements opposed to corporate globalization, specifically the activists grouped loosely around the International Forum against Globalization and the radical historians and political analysts of autonomist Marxism. Both of these two schools demonstrate important parallels between the first enclosures of the English feudal commons, which led to the grand transformation to capitalism and European imperialism and the continuing colonization and exploitation of shared resources throughout the world (Caffentzis 1995; Shiva 1993; Thompson 1968). They also show a historical line of succession from the creativity, resistance, and rebellion of the English commoners throughout the various colonies and diasporas of European colonialism to contemporary campaigns for local, democratic rule of shared resources (Dalla Costa 1995; Linebaugh and Rediker 2000; Shiva 2000).

The narrative of the English commons was one of protracted and often bloody struggle over land, a mode of production, a way of life, and over history itself. After the collapse of serfdom in the mid-fourteenth century, farm labor was in demand. A new class of small farmers, or yeomen, became responsible for much of the agricultural production, working the land under a complex system of open fields and common rights. They held customary right, or "copyhold," to a part of the feudal estate, as a sort of subtenancy. They also shared the use of untitled village land, marsh, and water holdings, in common with other small and medium-sized farmers and tenants (Travis 2000, 5). Some historians have called the fifteenth century a "golden age" of English labor, as many laborers were able to sustain themselves from their work on the land without needing to purchase additional commodities, while some were even able to accumulate wealth (Travis 2000, 5).

The English commons did not exist within a democratic society, but on the margins, or interstices, between state and private domains. Many rural families were poor and were subject to the domination of the feudal landlords via rents, levies, tributes, and taxes. However, there were many significant physical, social, cultural, and psychological times and spaces where the dominant classes and the commoners did not intersect. Open to all with a shared interest in their use, their value was derived from participation and was not a tradeable commodity (Shiva 1994). Not private, they were concerned more with continuing sustenance, security, and habitat, not with producing, distributing, or circulating commodities for a growth-oriented market system (Ecologist 1993). Commons regimes were also not public resources administered by the state, but instead were a form of direct rule by individuals and groups drawn from civil society, for the most part outside the electoral franchise.

As a corrective to the mainstream history, which argued that the commons were inefficient and unorganized, E. P. Thompson (1991, 131) documented their orderly use through a "rich variety of institutions and community sanctions which . . . effected restraints and stints upon use." Obligations, bonds, and evolving customary rights were defined and regulated as people negotiated multiple uses and schedules of space, time, labor power, and technical resources (Humphries 1990; Johnson 1996; Neeson 1993; Thompson 1968, 1991). This required the development of sophisticated interpersonal and community communication, which, in part, helps explain the origin of the words "communication" and "democracy" during this period. "Communication" meant "to make common to many," and democracy originated in the sixteenth century, when it meant "the rule of the comminaltie," the popular power of the multitude, implying the suppression of rule by the rich (Williams 1976, 93).

The scope and pace of the enclosures picked up in the fifteenth century and were an integral part of the grand transformation to European global capitalism (Thompson 1968). The first stage of the enclosures involved the fencing of common lands and copy-hold properties in order to introduce capital-intensive exploitation of the land for wool production. A new class of commercial landlords brutally dispossessed the rural population from grazing, fishing, hunting, quarrying, fuel, building materials, and rights of way. Eventually the landed gentry who dominated Parliament instituted enclosure laws.[5] The restructuring of the land and the way of life was instigated through a variety of measures, including engineering and highway projects, surveillance, the imposition of new work disciplines, systems of thought and governance, as well as dispersal and criminalization of all those who resisted. The enclosures, first developed in England, were extended to Ireland, Scotland, Wales, and then overseas, in the expropriation and exploitation of lands, waterways, and indigenous laborers throughout the Americas, Africa, and Asia.

However, the enclosures were not implemented without extraordinarily widespread and diverse resistance, a diversity of tactics that parallels those in use today. These ranged from moral and legal appeals, parliamentary petitions, and lobbying to fence breaking, arson and systematic trespass, and direct uprisings and riots. Some of the resisters articulated a radical communitarian philosophy. For instance, Gerard Winstanley of the Levellers argued that the common people should share equitably in the resources of the lands and waterways, negotiating their use among themselves without intervention by lords, military might, or parliamentary dictum.

The story of the commons provides insights into the different notions of democracy of the bourgeois and popular revolutionary traditions. Linebaugh and Rediker describe the saga of four hundred years of cross-Atlantic circulation of this heritage of commoners' creativity and revolt, tracing the ideas and experiments about popular rule and social justice of this motley crew to the French and American revolutions (Linebaugh and Rediker 2000). Vandana Shiva (2000) has described the legacy of women's direct action during the 1940s against the British Empire's privatization and extraction of rent from the land in contemporary Indian laws and democratic principles. Italian Marxist feminist Maria Rosa Dalla Costa links the Zapatista challenge to land enclosures brought about by the North American Free Trade Agreement (NAFTA) to the current campaigns against corporate globalization. "The webs of relations, analyses and information interweaving" among indigenous movements, workers, ecological movement militants, women's groups, and human rights activists are bridged by the continuing struggles for the commons, whose public spaces and ecology provide the possibility of "life, of beauty and continual discovery" (Dalla Costa 1995, 13).

The Internet and the Grand Transformation

We are now at the center of another grand transformation, from an economy dominated by industrial production to one in which information and digital knowledge play a key role in production, distribution, and circulation. One of the principal technologies of this new mode of production and social organization is the Internet. As the Internet has developed from a publicly funded network centered among universities, research institutions, and governments to one dominated by corporate commercial exchange, there has been a widespread debate over its ownership, governance, customary operation, model of communication, and relationship to democracy. There has also been a renewal of discussion and debate about the Internet as a new commons and new enclosure.

The Internet developed through both deliberate design and unintended consequence. An odd combination of social actors—U.S. military research, academic and corporate scientists, and grassroots social movements—used public resources and a high degree of creativity and collaboration to create this globally networked communications system.[6] The role of state and corporate players in the history of the Internet is of course much better known (Murphy, forthcoming). In response to the launch of the Russian *Sputnik* in 1957, the U.S. Department of Defense formed the Advanced Research Projects Agency (ARPA) to develop superior military technologies. They commissioned scientists in a number of different think tanks throughout the United States and western Europe. Working interdependently throughout the 1960s and 1970s, the scientists created the communications system that would become the Internet. They designed a decentralized system that allowed every node on the network to operate without centralized control centers, with the capacity to send and receive packets of digitally coded information.

This research work spawned a small number of research and development units in the 1970s near Stanford University in Palo Alto, California, which manufactured silicon chips (hence Silicon Valley) and personal computers. Silicon Valley provided the capacity for U.S., Japanese, and European corporations to "globalize" their production and distribution of goods and services. They developed an extensive network of computer-based Intranets, linked through the widely available and relatively cheap publicly regulated telephone lines. Foremost among these multinational corporations were the computer firms that moved their production to Asia, where strong U.S.-backed authoritarian regimes kept labor rates low.

During the 1990s, the dominant model of cyberspace shifted from publicly funded "information sharing" to a model of private commercial space (Menzies 2001, 219–220). The majority of users are no longer the .orgs and .govs who operated among decentralized communications networks of many to many. Much of the traffic is now dominated by the .coms and the broadcast model, in which a small number of dominant global media giants control the distribution pipelines into the Net, online traffic, and much of the content, exploiting this resource through fees, advertising, and subscriptions (Raphael 2001). However, this move to fence off or enclose the Net under corporate control has not been achieved without resistance. Among many tactics, Net users have lobbied and advocated with all levels of government and corporate actors, acted deliberately to break down barriers through sabotage or hacking, or refused to accept the privatization of common code and content through massive sharing of music and movie files (Dyer-Witheford 2002).

The New Commoners

The genesis of this resistance can be explored in the less celebrated history of the Internet and social movements, involving two principal sets of social movements (Murphy, forthcoming). The first group included the computer technicians who went on to develop many social communications uses for the Internet, to lobby for democratic policy, and/or develop hacking networks to fight the enclosure of cyberspace (Ludlow 2001). The next generation would become integral to the indymedia movement. The scientists and graduate students within the university research centers, and then the corporate factories of Silicon Valley, developed a communications system with the potential of allowing untold numbers of communicators to produce and distribute unlimited kinds and quantities of information with no central gatekeeping command. Operating with public money, they exchanged ideas to create software and hardware with open-source protocols that allow anyone to utilize and change the code. While some of these "geeks" or "techies" went on to become entrepreneurs, the development of the Internet, and especially the World Wide Web, owes much to this dispersed corps of individual techies, hackers, students, community-based organizations, and policy activists (Witheford 1997). By the 1990s, hundreds of individuals and groups, loosely collected in the open-source movement, were distributing information for free, sharing new software and hardware, and challenging the operating protocols of intellectual property through regulatory and entrepreneurial means. Another group was demonstrating the limits of existing corporate and state software and operating systems by sharing hacked software or warez (Pahati 2002).

This new class of knowledge workers operates in centers all over the world with a concept of collective intelligence in which they share a "common code" that is antithetical to proprietary ideas of intellectual property (Bosma et al. 1999; Castells 2001). While few would describe themselves as commoners, some speak in terms of breaking the corporate domination of the Internet and others think of themselves as contributing to democracy (Pahati 2002). Their mantra is that "information is free," that technology is a means to liberate information; their role is to allow information to circulate freely without the gate-keepers of nation-state or corporate domain (Castells 2001, 33). Regardless of their self-definitions, the open-source movement, the hackers, and file-sharing "pirates" have had a profound impact on the global Net, challenging the new corporate enclosures and attempting to keep the open architecture and free flow of information (Dyer-Witheford, 2002; Pfaffenberger 1999).

Building Networks against Corporate Globalization

Another group of social movements also identified with the communitarian aims of the early commoners. In the 1970s, a number of community-based groups began to use the new information technologies for social justice and social development. Several projects, such as Berkeley Community Memory, started in the Bay Area around Silicon Valley with the aim of making the information networks and communication capacity of the Internet publicly available. Other community-based organizations across North America and Europe developed a wide variety of new computer software and hardware for the Internet, including the Chicago group that developed the bulletin board system (BBS) and the movement of community nets to provide public access

(Castells 2001; Gutstein 1999).[7] By the 1980s, a number of international non-government organizations had realized the potential of linked international networks.

During the late 1980s, a coalition of national and international nongovernmental organizations (NGOs) from the northern and southern hemispheres acted to develop their own linked computer networks—including Geonet, Worknet, Fidonet, Econet, Greenet, Labornet, and Peacenet—allowing social movements of labor, ecology, peace, and women to share text-based information. This network of networks "preceded and long remained parallel to the commercialized Internet" (Murphy 2001, 7). In 1990, the Association for Progressive Communications (APC) was formed to support this global network, providing the first of many services with low-cost access to extensive resources at a global reach and speed, dramatically transforming the possibilities for political organization and action (Eagleton-Pierce 2001; Smith 2001).

Many of the NGOs had been working with social movements of small farmers, women, indigenous peoples, grassroots trade unionists, and environmentalists. There had been nationalist and left-wing critiques of the World Bank (WB) and International Monetary Fund (IMF) for their role in U.S.-driven capitalist exploitation of the Third World (Cleaver 1999). During the 1980s, this critique gained power as environmentalists and indigenous rights groups disseminated information about the impact of World Bank–funded megadevelopment projects on their livelihoods and cultures. Many NGOs allied to protest the imposition of World Bank and IMF policies of free trade and structural adjustment, which had privatized public resources, depleted public services, and weakened indigenous industries and national controls (Caffentzis 1995; Cleaver 1999; Dalla Costa 1995).

Some began to draw on the discourse of the commons. Vandana Shiva (1993, 215) argued that the the dominant ideology of postwar development focused on the enclosures of the national commons. Shiva (1994) and other ecologists framed debates over the uses of seeds and genetic materials as common regimes of knowledge and resources versus a corporate logic of enclosing and exploiting intellectual property.[8] *The Ecologist* magazine documented the efforts of a wide array of groups, outside the institutions of both market and state, to create or "defend open democratic community institutions that ensure people's control over their own lives" (Ecologist 1993, 175). Autonomist Marxist George Caffentzis described the privatization of the land tenure system in Africa as a "new enclosure movement" (Caffentzis 1995, 27). Maria Rosa Dalla Costa described the Zapatista revolt against NAFTA and the enclosure of commons lands as a struggle for the commons. However, the struggle was no longer local, or even national, but international in scope.

This new international movement continued to mobilize during the 1990s, as many of the NGOs and social-movement groups met face to face in international conferences and events. Some were counterconferences to multilateral organizations and strategems, including UN-sponsored meetings on the environment, women's rights, and human rights; WB, IMF, WTO, and NAFTA; the Asia Pacific Economic Conferences (APEC); and the Organization for Economic Cooperation and Development (OECD). Others, such as the *encuentros* in Mexico and Spain initiated by the Zapatistas, convened activists for exchanges and planning meetings and led to the formation of the People's Global Action (Moynihan and Solnit 2002). Seattle was to have been just one more international mobilization. However, the spotlight of the U.S. corporate media gave the movement a new level of power; Seattle was the coming-out party.

Seattle Independent Media Center

> The timing was right, there was a space, the platform was created, the Internet was being used, we could bypass the corporate media, we were using open publishing, we were using multimedia platforms. So those hadn't been available, and then there was the beginning of the anti-globalization movement in the United States. I think it was all of those pieces together. (Herndon 2001)

The roots of the IMC derive from these struggles over control of the resources of the cyber and terrestrial commons. The Seattle IMC brought together four sets of commoners: the social movements that were cooperating in "anti-globalization," local Seattle community activists, technicians from the open-source movement, and activist-media producers. While few would describe themselves as commoners, many used the discourse of commons and enclosures in their critiques of corporate privatization in general, and of the Internet and media gate-keeping in particular. They shared a vision of the IMC as an open, unbounded communications resource, whose "open publishing" innovation allows access to all. This new group of media workers are also like the early commoners, who operated their own copy-hold plot and shared the commons to sustain themselves without needing to buy commodities in the marketplace.

The indymedia commoners intend to be self-sufficient, volunteering their labor and supporting the local centers and the Net through a variety of grassroots efforts rather than depending on outside support. The Israeli site's banner, "You are your own journalist," and the Italian's "Don't Hate the Media—Become the Media," encapsulates the do-it-yourself approach. They see themselves as activists and journalists who produce their own firsthand accounts of campaigns in which they are involved and circulate the accounts of struggles from other sites all over the world.

The IMC in Seattle, and the international IMC movement, has also drawn on a legacy of organizational skills developed by earlier social movements (Herndon 2001). This process of sophisticated interpersonal and community communication is not unlike the earlier commons. One of the first things one will observe, on the Web sites and in face-to-face meetings, is the high level of democratic processing. The IMC network is based on a nonhierarchical structure that relies on highly complex processes of networked consensus. International meetings are held online. There are a wide array of listserv discussion groups that range from general discussions to finances to translation and technical issues. Meetings are conducted through highly complex processes of decision-making, using a consensus model drawn from the direct action wing of the anti-globalization movement.

Indymedia represents a new level of development of a communications commons. There had been earlier attempts among media activists to collaborate and share resources. Radical film documentarians in the

1960s and 1970s, the cable community-access movement of the 1970s and 1980s, Deep Dish TV and Free Speech TV via satellite TV, and micro-radio producers had all shared production and programming resources. Media-specific organizations of radio producers, video producers, and Web activists had been formed at the local, national, and international levels.[9] National and international conferences in San Francisco, New York, Amsterdam, Kuala Lumpur, and Delhi had convened activists interested in developing networks to promote media and democracy. However, many of these efforts had been stymied by the difficulties of sustaining long-term collaborations without stable financing, production facilities, or mechanisms for distribution; the craft separation into specific media technologies and practices; and rivalries for resources.

The Seattle IMC was able to surmount some of these barriers and move to a new level of social organization for a number of reasons. They carefully built a relationship with social-movement activists rather than distancing themselves from political organizing. They also consciously built on the experience of earlier networks, inviting many of the activists from the independent video, community radio, micro-radio, and open-source movements to participate very early on in the planning, fund-raising, and gathering of production equipment.[10] The storefront provided the personal and technological interface to bridge the rivalries between different media, different organizations, and different generations. More seasoned media activists worked together on production projects with newer producers and activists. The four-hundred-strong crew also used all the old and new media, from pens to laptops, and from inexpensive audio-tape and camcorders to the latest in digital recording technologies.

This high level of cooperation helped to break down, if not eliminate, some of the old craft and territorial divisions. Tom Poole, of Deep Dish TV, said: "In the early '90s, we all knew about each other but folks were more factionalized. Now you can see that there's a more collective effort" (quoted in Rinaldo 2000). During the WTO meetings in Doha, Qatar, in 2001, the IMC produced the "No New Rounds" radio Webcast offshore, together with Greenpeace; and in 2002, with "Democracy Now," the IMC broadcast from the protests against the World Economic

Forum in New York City and the World Social Forum events in Porte Alegre, Brazil. After the World Social Forum, a caravan of media activists from several different groups covered the crisis in Argentina, reporting directly from the mass meetings in the streets.

The success of the IMC was also due to the new array of available digital technologies. As a high-tech center, Seattle was also home to the original technical support crew, and the technical crew remains an indispensable part of the IMC. Most of the Centers still operate on the same donated ISP and use open-source software. The IMC also took advantage of the advancements in digital video. The new lightweight digital cameras are cheap, easy to work with, and edit and can broadcast instantaneously, allowing much more collaboration. Eric Galatas from Free Speech TV thinks that television will change dramatically as a result. "There are so many people now picking up DV [digital video] cameras, getting their hands on iMacs or G4s and editing great videos. . . . I think we're going to look back on this period as a launch pad for an entirely new way of making and distributing television" (quoted in Rinaldo 2000).

Most important, the IMC could overcome the limited space and the distribution problems inherent in the old media. The Internet and related technologies enabled a quantum leap in time and space for other kinds of content generation as the site could accept an unlimited amount of content, including text, photos, graphics, video, and audio. While debates over how to sustain the resource continue, there are none of the space limitations, and ensuing conflicts over sharing, that led to the crisis at Pacifica and constant tensions among other older independent media. Also, the reach is potentially so much further: During the anti-WTO protests in Seattle, the site had a million and a half hits, and the entire network is now estimated to receive about four-hundred thousand page views a day (Pavis 2002).

The IMC represents a major step forward in the tactical use of autonomous media. It has brought together activists and journalists from across the different media with movements that were able to circulate their messages on a scope and scale not realized before. In many ways, they have been able to surmount the limits on the resource that always faced the land-based commons and earlier media commons. The expanded horizon for production and distribution has limited the battles over resources, but not eliminated them. The IMC is networked; highly consultative decision-making owes a lot to skills developed in the consensus-model training of the direct action wing of the antiglobalization movement. The negotiation of resources appears to operate with far fewer of the stand-offs that seemed inherent in the earlier activist media movements of which I was a part. However, in the long-term, some of the same old questions remain.

How can the IMCs sustain this resource? The decentralized network model helped share the labor and the fund-raising. However, the dependence on volunteers and the sharing of a limited number of resources will be hard to continue indefinitely. Already, those people who are able to volunteer tend to represent a small minority of young white North Americans and Europeans who can afford to share their time (Rinaldo 2000). The network is facing these problems in creative ways, sending the caravan to Argentina, sending volunteers with technical expertise to new sites in Latin America, providing constant technical and other kinds of support via the Internet itself, and circulating key personnel through the network. Nevertheless, very creative solutions are needed to overcome the huge inequality of access to media production and Internet technologies that exists among working-class communities of color in North America and Europe, and even more so in the southern hemisphere.

The success of the IMC network has not been without other challenges and costs. Its visibility has brought more attention from national and international security agencies. In Seattle, the IMC had been able to operate as witnesses, providing a thin skin of protection against greater police violence and a photographic and audio record for the legal teams fighting police actions. However, after the confrontation between demonstrators and police during the spring 2001 Free Trade Agreement of the Americas (FTAA) meeting in Quebec City, the Seattle site was raided by the FBI, based upon information from the Canadian Security Intelligence Service (CSIS). During the 2001 Genoa Meeting of the G8 in Genoa, Italy, the Italian police attacked the IMC, beating and arresting everyone inside (Halleck 2002; Starhawk 2001). What are the risks of more security intervention in IMC offices and Web sites?

Among the monocultural enclosures of the .coms and media giants, indymedia is a vibrant commons. The IMC produces counterinformation to the media giants, and are able to do so using the same communication and information machinery that capital uses to ensure its own mobility (Witheford 1997, 205). The IMC has built a network from the heritage created by earlier media activists and, as importantly, has based itself within the social movements against corporate globalization, acting to make visible and circulate a multiplicity of social movements and actions.

Notes

1. A banner on the homepage of the Italian Independent Media Centre, September 2001.

2. The local daily and weekly Seattle newspapers presented a range of views about the impact of WTO free trade decisions on the environment, labor standards, and local democratic governance. However, the initial television coverage and national mainstream coverage focused on the few incidents of property damage; characterized the wide range of protesting groups as laughable and ill informed, and dismissed their critiques of the WTO and corporate globalization. One *New York Times* columnist summed up this trope about demonstrators: "a Noah's ark of flat-earth advocates, protectionist trade unions, and yuppies looking for their 1960s fix" (FAIR Media Advisory, December 7, 1999). A 2001 study in the *Columbia Journalism Review* shows how this framing leans "heavily towards the corporate side in major news organizations coverage of protests at the IMF meeting in Prague, the FTAA talks in Quebec City, the European Summit in Sweden, and the G-8 meeting in Genoa" (cited in Hyde 2002).

3. In Kidd 1998 I teased out the concept of the radio communications commons, based in the electrospace.

4. Hannibal Travis (2000) argues that enclosure includes the actual process of enclosure, and as importantly, a "recharacterization of existing entitlements as theft" (4).

5. The Houses of Parliament who enacted the Enclosure Laws were dominated by the landed gentry and noblemen in the top 1 percent income bracket (Travis 2000, 5).

6. There are many good histories of the Internet that show this complex development, propelled by very different social actors (Bosma et al. 1999; Castells 2001; Dyer-Witheford 1999; Murphy 2001).

7. This development of public use software and hardware continues throughout the world. Most recently, Indian and Brazilian computer designers have

developed cheap personal computers for mass use and have adapted open-source software for operating systems.

8. In the plenary address to the counterconference against the World Economic Forum in Melbourne, Vandana Shiva compared the struggle for farmers' control of their seed to the campaign for free computer software (Shiva 2000, 9).

9. The Newsreel Organization built national and international links among radical film producers. International video producers had convened in conferences in the 1980s and 1990s under the loose direction of Videazimut. Grassroots radio producers formed an international organization called the World Community Radio Organization (AMARC), which also facilitated collaborations among feminist and indigenous producers. The Tactical Media Conferences in Amsterdam convened activists from old and new media (Bosma et al. 1999; Halleck 2002; Kidd 1998).

10. Collaborators included Deep Dish TV, Paper Tiger TV; Free Speech TV; Whispered Media; Changing America; NY Free Media Alliance; the micro-radio producers, including Free Radio Berkeley and Prometheus Media; as well as many others.

References

Barstow, Anne Llewellyn. 1994. *Witchcraze: A New History of the European Witch Hunts*. San Francisco: Pandora Books.

Basuki, Tedjabayu. 1999. "Indonesia: The Web as a Weapon," in The Next Five Minutes 3 Public Debates List. Online. http://www.n5m.org.

Bosma, Josephine, Pauline van Mourik Broekman, Ted Byfield, Mathew Fuller, Geert Lovink, Diana McCarty, Pit Schultz, Felix Stalder, McKenzie Wark, and Faith Wilding. 1999. *Read Me! Filtered by Net Time: ASCII Culture and the Revenge of Knowledge*. Brooklyn, NY: Autonomedia.

Caffentzis, George. 1995. "The Fundamental Implications of the Debt Crisis for Social Reproduction in Africa," pp. 42–57 in *Paying the Price: Women and the Politics of International Economic Strategy*, ed. Maríarosa Dalla Costa and Giovanna Dalla Costa. London: Zed Books.

Castells, Manuel. 2000. *The Rise of the Network Society*. Oxford: Blackwell Publishers.

——. 2001. *The Internet Galaxy: Reflections on the Internet, Business and Society*. Oxford: Oxford University Press.

Cleaver, Harry. 1995. "The Electronic Fabric of Struggle." Online. http://www.eco.utexas.edu/faculty/Cleaver/zaps.html.

——. 1999. "Computer-Linked Social Movements and the Global Threat to Capitalism." Online. http://www.eco.utexas.edu/faculty/Cleaver/hmchtml-papers.html.

Curtin, Michael, and Thomas Streeter. 2001. "Media," pp. 225–249 in *Culture Works: The Political Economy of Culture*, ed. Richard Maxwell. Minneapolis: University of Minnesota Press.

Dalla Costa, Maria. 1995. "Development and Reproduction." *Common Sense* 17: 11–33.

Dyer-Witheford, Nick. 1999. *Cyber-Marx: Cycles and Circuits of Struggle in High-Technology Capitalism*. Urbana: University of Illinois Press.

——. 2002. "E-Capital and the Many-Headed Hydra," in *Critical Perspectives on the Internet*, ed. Greg Elmer. Lantham, MD: Rowan and Littlefield.

Eagleton-Pierce, Matthew. 2001, September. "The Internet and the Seattle WTO Protests." *Peace Review*. 13(3): 331–38. Special Issue: Social Justice Movements and the Internet.

Ecologist, The. 1993. *Whose Common Future? Reclaiming the Commons*. Philadelphia and Gabriola Island, B.C.: New Society Publishers.

Esteva, Gustavo. 1993. "Development, " pp. 6–25 in *The Development Dictionary: A Guide to Knowledge and Power*, ed. Wolfgang Sachs. London: Zed Books.

Ford, Tamara Villarreal, and Genève Gil. 2001. "Radical Internet Use," pp. 201–34 in *Radical Media: Rebellious Communication and Social Movements*, ed. John Downing et al. Thousand Oaks, CA: Sage.

Gutstein, Donald. 1999. *e.con: How the Internet Undermines Democracy*. Toronto: Stoddart.

Halleck, DeeDee. 2002. *Hand-Held Visions*. New York: Fordham University Press.

Hardt, Michael, and Antonio Negri. 2000. *Empire*. Cambridge, MA: Harvard University Press.

Herndon, Sheri. 2001, July 20. Telephone interview.

Humphries, Jane. 1990. "Enclosures, Common Rights, and Women: The Proletarianization of Families in the Late Eighteenth and Early Nineteenth Centuries." *Journal of Economic History* 50(1): 17–42.

Hyde, Gene. 2002. "Independent Media Centers: Cyber-Subversion and the Alternative Press." *First Monday*. 7(4). Online. April. http://firstmonday.org/issues/issues_4/hyde/index.html

Independent Media Center brochure. 2001. Seattle, WA: Independent Media Center.

Johnson, Mathew. 1996. *An Archaeology of Capitalism*. Cambridge: Blackwell Publishers.

Kidd, Dorothy. 1998. "Talking the Walk: The Media Enclosures and the Communications Commons." Doctoral Dissertation. Simon Fraser University.

——. 2001, September. "Introduction." *Peace Review* 13(3): 325–30. Special Issue: Social Justice Movements and the Internet.

Kidd, Dorothy, and Nick Witheford. 1994, November 12. "Counterplanning from Cyberspace and Videoland: or Luddites on Monday and Friday, Cyberpunks the Rest of the Week." Paper presented at "Monopolies of Knowledge: A Conference Honoring the Work of Harold Innis," Vancouver.

Land, Jeff. 1999. *Active Radio: Pacifica's Brash Experiment.* Minneapolis: University of Minnesota Press.

Lasar, Mathew. 1999. *Pacifica Radio: The Rise of an Alternative Network.* Philadelphia: Temple University Press.

Lessing, Lawrence. 1999. *Code and Other Laws of Cyberspace.* New York: Basic Books.

——. 2001. *The Future of Ideas: The Fate of the Commons in a Connected World.* New York: Random House.

Linebaugh, Peter, and Marcus Rediker. 2000. *The Many-Headed Hydra: Sailors, Slaves, Commoners, and the Hidden History of the Revolutionary Atlantic.* Boston: Beacon Press.

Ludlow, Peter, ed. 2001. *Crypto Anarchy, Cyberstates, and Pirate Utopias.* Boston: MIT Press.

Menzies, Heather. 2001. "On Digital Public Space and the Real Tragedy of the Commons," pp. 217–28 in *e-commerce vs. e-commons: Communications in the Public Interest*, ed. Marita Moll and Leslie Regan-Shade. Ottawa: Canadian Centre for Policy Alternatives.

Messman, Terry. 2001. "Justice Journalism: Journalist as Agent for Social Change." *MediaFile* 20(4): 1.

Midnight Notes Collective. 1992. *Midnight Oil: Work, Energy, War 1973–1992.* Brooklyn, NY: Autonomedia.

Moynihan, Denis, and David Solnit. 2002. "From the Salt Marshes to Seattle: Direct Action's History," pp. 129–34 in *The Global Activist's Manual: Local Ways to Change the World*, ed. Mike Prokosch and Laura Raymond. New York: Thunder's Mouth Press/Nation Books.

Murphy, Brian Martin. 2001. "Propagating Alternative Journalism through Social Movement Cyberspace: The Appropriation of Computer Networks for Alternative Media Development," in *Appropriating Technology: Vernacular Science and Social Power*, ed. R. Eglash, J. Croissant, G. DiChiro and R. Fouché. Minneapolis: University of Minnesota Press.

——. 2002. "Towards a Critical History of the Internet." *Critical Perspectives on the Internet*, ed. Greg Elmer. Lanham, MD: Rowan and Littlefield.

Neeson, J. M. 1993. *Commoners: Common Right, Enclosure and Social Change in England, 1700–1820.* Cambridge: Cambridge University Press.

Pahati, Omar. 2002. "Digital Pirates and the "Warez' Wars." Online. www. alternet.org (January 24).

Paton, Dean. 1999. "War of the Words: Virtual Media versus Mainstream Press." Online. www.csmonitor.com (December 3).

Pavis, Theta. 2002. "Modern Day Muckrakers: The Rise of the Independent Media Center movement." Online. http://www.ojr.org/ojr/business/1017866594.php. *OnLine Journalism Review*. USC Annenberg.

Pfaffenberger, Bryan. 1999, December 13. "In Seattle's Aftermath: Linux, Independent Media, and the Survival of Democracy." www.linuxjournal.com/article.php?sid=5075 (December 13).

Project for Excellence in Journalism. 2001. "Before and After: How the War on Terrorism Has Changed the News Agenda." Network Television. June–October. Columbia University School of Journalism. Pp. 1–21.

Raphael, Chad. 2001. "The Web," pp. 209–10 in *Culture Works: The Political Economy of Culture*, ed. Richard Maxwell. Minneapolis: University of Minnesota Press.

Regan, Tom. 1999. "News You Can Use from the Little Guys." *Christian Science Monitor*. Online. www.csmonitor.com (December 9).

Rifkin, Jeremy. 2000. *The Age of Access: The New Culture of Hypercapitalism Where All of Life Is a Paid-For Experience*. New York: Jeremy P. Tarcher/Putnam.

Rinaldo, Rachel. 2001. "Pixel Visions: the Resurgence of Video Activism." Online. www.libmagazine.org/articles/featrinaldo_115_p.htm. (July 9).

Ryan, Charlotte, Kevin M. Carragee, and William Meinhofer. 2001. "Theory into Practice: Framing, the News Media, and Collective Action." *Journal of Broadcasting and Electronic Media* 45: 175.

Sénécal, Michel. 1991. "The Alternative in Search of Its Identity," pp. 209–218 in *Video the Changing World*, ed. Nancy Thede and Alain Ambrosi. Montreal: Black Rose Books.

Shiva, Vandana. 1993. "Resources," pp. 206–18 in *The Development Dictionary; A Guide to Knowledge and Power*, ed. Wolfgang Sachs. London: Zed Books.

——. 1994. "The Recovery of the Commons, A Public Lecture." Alternative Radio Production.

——. 2000. Plenary address to RMIT University conference "Global Capital, Local Responses," Australian Broadcasting Corporation, Melbourne, Australia.

Smith, Jackie. 2001, Spring. "Cyber Subversion in the Information Economy." *Dissent*: 48–52.

Solomon, Norman. 2001. "When Journalists report for Duty." Online. http://www.fair.org/media-beat/010920.html.

Starhawk. 2001. "Lifelong Activism: Finding the Courage, Tenacity, and Love," pp. 262–264 in *Global Uprising: Confronting the Tyranies of the 21st Century*, ed. Neva Welton and Linda Wolk. Gabriola Island, B.C.: New Society Publishers.

Tarleton, John. 2000, Winter. "Protesters Develop Their Own Global Internet News Service." *Mark Nieman Reports*, 54(4): 53–55.

Thompson, E. P. 1968. *The Making of the English Working Class*. Harmondsworth, England: Penguin Books.

——. 1991. *Customs in Common*. London: Merlin Press.

Travis, Hannibal. 2000, Spring. "Pirates of the Information Infrastructure: Blackstonian Copyright and the First Amendment." *Berkeley Technology Law Journal*, 15(2): 777.

Williams, Raymond. 1976. *Keywords: A Vocabulary of Culture and Society*. New York: Oxford University Press.

Witheford, Nick. 1997. "Cycles and Circuits of Struggle in High-Technology Capitalism," pp. 195–242 in *Cutting Edge: Technology, Information, Capitalism and Social Revolution*, ed. Jim Davis, Thomas Hirschl, and Michael Stack. London: Verso.

3

Classifying Forms of Online Activism

The Case of Cyberprotests against the World Bank

Sandor Vegh

Introduction

This chapter classifies political activism using the Internet. It is part of a larger work that situates online activism in a power struggle of control and resistance between the power elite and the public. First, I analyze the forms of online activism and construct a model to examine particular movements in terms of their utilization of the Internet. I cite examples to show the diversity of these activities, regardless of their political cause or geographic location, and provide a system for understanding various forms of online activism. Second, I examine how the online aspects of the anti-globalization protests against the World Bank fall into this classification system. I pay particular attention to the forms and results of online dissent preferred by the activists. Finally, I examine the strategies the Bank deploys to handle these cyberprotests.

I define online activism as a politically motivated movement relying on the Internet. The scenario is fairly simple: Activists now take advantage of the technologies and techniques offered by the Internet to achieve their traditional goals. Their strategies are either Internet-enhanced or Internet-based. In the former case, the Internet is only used to enhance the traditional advocacy techniques, for example, as an additional communication channel, by raising awareness beyond the scope possible before the Internet, or by coordinating action more efficiently. In the latter case, the Internet is used for activities that are only possible online, like a virtual sit-in or hacking into target Web sites. Online activism is comprised of proactive actions to achieve a certain goal or of reactive actions against controls and the authorities imposing them. To better understand the forms, impact, and goals of online activism, I now turn to its different forms for categorization.

At first glance, the types of Internet activism fall into three general areas: awareness/advocacy; organization/mobilization; and action/reaction. This typology emphasizes the direction of initiative—whether one sends out information or receives it, calls for action or is called upon, or initiates an action or reacts to one. These are progressive steps of online activism leading from basic information seeking and distribution to online direct action, better known as "hacktivism."

Awareness/Advocacy

Public awareness is achieved by accessing information that is relevant to the cause. Naturally there is often difficulty involved. Since the traditional information channels may well be controlled by those whose interest is counter to that of the activists, the Internet may serve as an alternative news and information source. The news and information are provided by individuals and independent organizations, largely focusing on events and issues not reported, underreported, or misreported in the mainstream mass media. The forms of obtaining information include visiting relevant Web sites or participating in different types of e-mail distribution lists.

Information distribution on the Internet has another important implication for activism. It creates distribution networks that can later be used for organization and mobilization purposes. For instance, Tedjabayu (1999) reports on the NusaNet Consortium—a restricted and Pretty Good Privacy (PGP)-encrypted inter-NGO e-mail system—created to disseminate alternative news among themselves and to the larger population. Especially when a serious violation of human rights occurs, the Internet is essential in reporting the atrocity to the outside world, attracting public condemnation and fueling subsequent action.[1] It is usually a particular incident—a conflict—that triggers and fuels an activist movement aided all along by the Internet. These networks being in place proves extremely useful when the moment for political change arrives.

These information-distribution networks are also prevalent in the forms of dissident communities online. The role of these communities (operating mostly in the form of listservs, Usenet groups, or discussion groups) is twofold. First, they provide a channel into their nondemocratic home country by sending in news that is otherwise banned there. Take, for example, the apakabar listserv based in Maryland dealing with Indonesia, the BurmaNet list, or the China News Digest. Second, they provide a forum for open discussion on censorship or human rights violations occurring within closed authoritarian regimes. Often, news from inside the regime finds its way to these forums where it is acted upon in forms of protests, boycotts, or lobbying. In many cases, dissidents have attempted to influence domestic politics through the mass media, but they have made little impact since foreign news sources are usually banned in these countries. However, there is a more indirect way to exert pressure, using commercial means to achieve political goals. Although it belongs to the next category of resistance activities, it is appropriate to cite here the example of Zar Ni, a student in Wisconsin who started the Free Burma Coalition. This coalition instigated a massive campaign against the Burmese government and exerted enough pressure on transnational corporations that they cut their investments in Burma.[2] Ni's group also managed to raise enough public and political awareness of the situation in Burma that it was placed considerably higher on the U.S. foreign policy agenda.

The primary uses of the Internet in online advocacy revolve around organizing the movement and carrying out action. The actors can be either part of a strictly defined group (e.g., an NGO), a civic advocacy group, a lobbying body, or a loosely defined group (e.g., the anti-globalization protesters or the Zapatista activists). Similarly, the process of online advocacy can focus on organizing and mobilizing a group of people for action, or actually carrying out an effort with a particular goal in mind. For example, in the anti-globalization protests the Internet is used mainly for coordinating action. The Internet enables the large number of activist groups and individual protesters involved (tens of thousands in most cases) to establish a time- and cost-efficient communication channel. Only the Internet allows an activist to distribute a message to thousands of people all over the world at once and to publish information that is accessible from anywhere anytime with virtually no cost. Protestors' conscious and efficient use of the Internet is exemplified by the centralized Web site and e-mail distribution list that is set up for each major protest to bring together the scores of participating activist organizations, coordinate their actions, and provide practical information ranging from accommodation and places to eat cheaply to methods of nonviolent resistance against police brutality.

Lobbying is a more traditional form of advocacy. It also has three different types distinguished by the target of action. The first and most traditional type is aimed at one's own government's legislative body. For example, companies have been formed specifically to provide cyberlobbying on behalf of individuals and organizations less prepared to undertake a successful online campaign (e.g., e-Advocates). In another example, coalitions formed around a particular issue have been involved in online lobbying. The possibility of the second type of lobbying—influencing worldwide opinion—owes a great deal to the Internet. For example, it probably would not have been possible to familiarize the world with the case of Chiapas in Mexico had the Zapatista movement not relied on the Internet for communication and mobilization. The third type of lobbying targets the government of oppressive regimes. It has been shown that undemocratic regimes are now more responsive to the Internet in their propaganda efforts. Yet this tactic is still more likely to yield results if pressure is applied indirectly, as demonstrated by the Free Burma Coalition campaign.

Organization/Mobilization

The Internet is used for mobilization in three different ways. First, it can be used to call for offline action, as exemplified by a distributed e-mail or posted Web site that calls for a demonstration at a given place and time. Second, it can be used to call for an action that normally happens offline, but can be more efficiently done online, such as a call for contacting one's congressional representative through e-mail. The efficiency lies in the minimal time it requires to compile a message especially if templates are provided. Whether the resulting considerably larger number of electronic messages makes the same impact on the legislator as hand-written letters is another question. Third, the Internet can be used to call for an online action that can only possibly be carried out online, such as a coordinated massive spamming campaign or ping-storm attack, which maliciously saturates a server with messages aimed to test communication between computers in an amount and frequency that overwhelm its response capacity and therefore disrupts or halts the server of the target entity. This last type, however, falls into the next category of online activities.

The technical aspects behind online advocacy and mobilization are fairly straightforward. The most effective way is to set up a Web site that provides information and influences the readers to adopt the desired point of view and prompts them to take action on the side of the cause. In parallel, e-mail lists must be set up to provide a forum for the larger public to discuss the issues, to distribute news and development on the issue for people already sympathetic to the cause, and to enable communication and coordination among those in the core of the movement. Certainly, these discussion forums may take the form of any online synchronous or asynchronous communication technology. The key is, as the e-Advocates say, to match the online tools to the task (Fielding and Bennett 2001). Moreover, the most successful online advocacy campaigns seem to be the ones that combine the different types of lobbying and mobilization.

Action/Reaction

In a very simplistic, media-instigated view the last category covers online attacks committed by "hackers." Of course, this sentence in itself summarizes the problem with the popular understanding of this more proactive and aggressive use of the Internet to achieve a goal that can be both politically and financially motivated. I describe trends in order to demonstrate the diverse manifestations of online activism and the ambiguous terminologies popularized by the mass media. (The colloquial and media usage of the word "hacker" and the subsequent popular image of the hacking community are discussed elsewhere; see Himanen 2001; Levy 1984; Sterling 1992; Taylor 1999.)

Perhaps the first well-known example of hacktivism was the pro-Zapatista movement in support of the struggle of the indigenous people of Chiapas against the oppression of the Mexican government. The support network formed around the Zapatistas helped the worldwide distribution of the communiqués of their leader, Subcomandante Marcos, over the Internet. It is important to note that the Internet played a role only in the external communication of the movement with the help of activists in countries with more advanced communication capabilities. Despite the media-instigated allusions, the image of Subcomandante Marcos typing his communiqués on a laptop hooked to the Internet over a satellite connection in the middle of the jungle is rather a myth. Similarly, Osama bin Laden is not likely to be sending out his evil plots to other terrorists from the middle of the Afghan desert on a wireless portable computer. Thus, the Internet-based part of democratizing (or undemocratizing) movements is usually provided by sympathizers located in more technologically advanced countries.

In the case of the Zapatistas, the main support comes from the Electronic Disturbance Theatre (EDT), a group of activists and artists engaged in practicing what they call "electronic civil disobedience." Their activism focuses on the coordination of selective direct action against anti-Zapatista entities, such as the Mexican and U.S. governments, and financial institutions in Mexico City. It is aimed at overwhelming the target Web servers with requests, radically slowing or shutting them down. In order to automate this coordinated attack they created an application called FloodNet, which was released to the public and came to be widely used in other online direct actions. While overwhelming servers with requests is hardly a crime, any direct action that results in disrupting the operation of such servers may, in fact, constitute a legally actionable activity. In accord with their name and philosophy, however, the EDT people see their actions differently: "We considered it performance art. . . . When we do a performance, our performance or our actions are considered symbolic gestures; we are trying to bring attention to a particular event or cause—we are not trying to do any criminal activity. We just don't want to do that" (Conan 2001).

Similarly, the Internet plays an ever-increasing role in the coordination of the global resistance against capitalist imperialism. Apart from organization and information distribution, activists set up a fake Web site for the World Trade Organization (WTO) to further their own agendas, www.gatt.org, fooling many participants into believing that they had actually visited the real conference site. They also tried to hijack traffic to the official www.wtoseattle.org Web site by creating their version cunningly named as www.seattlewto.org.[3] In addition, the Internet was also used to launch an attack online against the WTO Web site in parallel with the street protests at their meeting in November 1999 in Seattle. The party responsible for the hack that disabled the WTO servers was the Electrohippies Collective, a group of campaigners, computer specialists, and consultants based in Great Britain.

When the WTO attempted to shut down the www.gatt.org site, an activist group called The Yes Men responded by releasing Reamweaver, an automatic Web site parody software. Concurrently with the World Economic Forum meeting a few months later in New York, the group "reamed" the official www.weforum.org site—which also fell victim to Denial of Service (DoS) attacks—producing the www.world-economic-forum.com parody site. Their hacktivism software, just like EDT's FloodNet applet, was funded by ®TMark, a collective that supports genuine and ingenious ways to sabotage corporate culture by channeling funds from donors to workers who carry out these projects. It operates like any incorporated entity, mimicking the world of mutual funds, investments, and stocks.

Defacing Web sites and disrupting servers are not the means of hacktivism for every politically minded hacker. Self-described as the most influential group of hackers in the world, the Cult of the Dead Cow (cDc) operates on a different philosophy. They define hacktivism as "the use of technology *to advance human rights* through electronic media" (cDc 2001, italics mine). They firmly believe in freedom of information and freedom to communicate. Accordingly, their actions are targeted against authorities censoring or controlling the Internet. Their latest initiative is the "peekabooty" project, a distributed collaborative privacy network that allows for accessing DNS-filtered content through participating servers by users living in countries that censor the Internet. In a way, the cDc is fighting government censorship online. Their activism links politicized hackers, human rights activist groups, and state-level agendas together.

Although often implicated almost to the point of certainty, so far governments have not been directly linked to cyberattacks. These acts are usually carried out by individuals or groups, with or without their governments' knowledge or consent. This cyberpartisanship may be seen internally as a civil response to state affairs, and externally as nonstate action supporting the state-led political agenda or military engagement. Citizens responded on the domestic level to unpopular regulations, violations of minority rights, or outright censorship by targeting their own government's information systems (e.g., Burma, China, Mexico, Sri Lanka). In another scenario of internal conflict, Internet attacks are part of independence movements by activists targeting computers of the government that claims authority over them (e.g., China/Taiwan, Indonesia/East Timor, Israel/Palestine, Yugoslavia/Kosovo). On the international front, hackers attacked computers of foreign governments with which their countries were at peace (e.g., China/Indonesia, China/Japan, China/United States of America, Hungary/Romania), in occasional military conflict (India/Pakistan), or at war (United States of America/Yugoslavia). Each of these pairs of countries provides for an excellent case study.

One of the first documented cyberattacks occurred in Sri Lanka in 1997 in support of the Tamil Tiger separatists to disrupt government communications by overloading Sri Lankan embassies with thousands of e-mails. In addition to e-mail boxes, Web sites are the popular targets for attack. As part of the campaign for the full autonomy of East Timor, about four dozen Web sites were defaced by the "KaotiK Team" in a collective hacking on August 1, 1998. But perhaps the most "real" cyberwar occurred on January 28, 1999, when hackers brought down the East Timor virtual country domain, www.freedom.tp. Forces sympathetic to the Indonesian government were blamed for the cyberattack. East Timor declared its virtual independence on the twenty-second anniversary of Indonesia taking over the province on December 7, 1997, with the creation of its own top-level domain (.tp) and the launching of the freedom.tp site two days later. It was served from an Irish Internet service provider, Connect-Ireland.[4]

On the outskirts of the People's Republic of China a more serious cyberconflict has been under way for years between the mainland and Taiwan. While the former is undoubtedly superior in traditional military power, the tiny island nation dominates in information technologies. Soon after the Tiananmen Square demonstrations a computer virus called "Bloody" (or 6/4) hit mainland China. This is only one of the many originating from Taiwan; the rest included such infamous viruses as Michelangelo and Chernobyl. Chinese governmental computers were also hacked for human rights and freedom of speech abuses. In 1998, a human rights Web site set up by the Chinese government was hacked and redirected to Amnesty International; governmental computers blocking access to blacklisted sites were also targeted. On the other side, Chinese hackers launched massive attacks against Taiwanese government Web sites following the president's comments that disaccorded with the "One China" policy.

In another instance, Indonesian Web sites were hacked as a response to Indonesia's treatment of its ethnic Chinese population. Similarly, several dozen Japanese government sites came under a collective massive attack following Japanese politicians' speeches denying the 1937 Nanking Massacre. In 1999, after the accidental NATO bombing of the Chinese Embassy in Belgrade, hackers from China launched attacks against U.S. government information systems. These cyberoffensives included the full range of e-warfare devices including replacing official Web pages with protest material and offensive language, postings in chat rooms and newsgroups, e-mail spamming aimed to disrupt or shut down networks, and DoS attacks.

According to an iDefense white paper, hackers defaced at least 330 sites during the month of April 2001, 316 of which having .cn top-level domains (iDefense 2001b). The defacement spree, referred to as the "Labor Day Strike," was summarized as an online response to the diplomatic tensions between the United States and China following the collision of a U.S. surveillance plane and a Chinese fighter jet on April 1, 2001. The cyberconflict peaked on May 3, when the Chinese hacker alliance committed some three hundred defacements that day only. A truce was eventually declared on May 9, after the unleashing of an anti-American defacing worm. It is not clear, however, who exactly wrote the worm that spread as e-mail attachments and infected machines, which then sought out vulnerable Web servers and defaced the main Web sites with an anti-American message. By then, the pro-Chinese defacements totaled over one thousand during the previous few weeks.

Apart from particular events that prompt activism online on both sides, territorial disagreements are often in the center of hacktivist raids. While the question of Transylvania has been in the Hungarian and Romanian national consciousness since its controversial cession to Romania in 1920, it finally surfaced in cyberspace in the form of a now yearlong hacker war between the two countries. It apparently started with Romanian hackers collectively targeting Hungarian revisionist Web sites. The counterstrike, which followed soon after, attacked sites of the Romanian right-wing movement. The cyberclash culminated in the spectacular defacement of the Office of Hungarian Minorities Abroad Web site on February 15, 2002. What puzzled many regarding the perpetrators, however, was that the Romanian tricolor national banner was placed on the page with the colors in the wrong order.

The territorial dispute of India and Pakistan over the Kashmir region illustrates propaganda war and Web site hacking between countries in occasional military conflict. The Indian Army's Kashmir Web site launched in September 1998 as counter-propaganda to other sites supporting Muslim Kashmiris seeking independence was "hijacked" allegedly by hackers from Lahore, Pakistan.

Another widely publicized cyberconflict between Israeli and Palestinian hackers also demonstrated a range of attack tools, including defacements, e-mail bombs, ping storms, and distributed DoS attacks. This hacker war has been going on for years now, complementing the deadly reality of the Middle East conflict to this day. As a notable development, pro-Palestinian attack sites made available computer viruses (LoveLetter, CIH, and Melissa) along with Word macro viruses for use against Israeli sites, which marked the first confirmed distribution and strategic use of viruses in a cyberconflict (iDefense 2001a). The attack strategies also showed sophistication, in a well-planned, multiphased, coordinated attack against carefully selected target sites.

The use of the Internet during a de facto war (American aggression or NATO peacekeeping mission, depending on the political views), however, is best exemplified by the Yugoslav conflict. Serbian and Kosovar hackers fought their own cyberwar online. Crna Ruka (Black Hand), Yugoslavia's hacker team, and the Kosovo Hackers Group attacked and defaced the enemy's and each other's online news sites. E-mail accounts of U.S. media companies, political decision-makers, and academics were flooded with messages from Yugoslavia in an e-campaign that came to be called "YugoSpam." Cyberattacks were launched against NATO and Pentagon computers, and the U.S. Navy Web server was hacked, while the official NATO server was brought down by a ping storm.

This cyberwar had the potential of escalating to new, unknown heights, but the United States backed off from an electronic counter-strike. "During the Kosovo war, U.S. officials are reported to have decided against deploying their electronic arsenal because of fears that the impact on civilian life would have led to charges of war crimes under the Geneva convention" (Havely 2000). Regardless, the danger was imminent; as Richard Clarke, National Security Council terrorist coordinator, explained: "An attack on American cyberspace is an attack on the United States just as much as a landing on New Jersey. The notion that we could respond with military force against a cyberattack has to be accepted" (Miklaszewski and Windrem 1999).

As these examples demonstrate, the sustained acts of hacktivism are often labeled as cyberwar. The colloquial use of the constituent words is one source of the seemingly interchangeable nature of these concepts. Furthermore, the use and connotation of the terms also vary with the actual source and perspective (e.g., military, media, political activists, or the hacking community). For example, some civil activists or academicians engaged in or studying online activism either regard it as a distinct terrain of the military or attempt to connect it to the field of cyberwarfare (Denning 1999; Wray 1998). On the other hand, the military either tries to force online activism into its logic of thinking (substate or nonstate actors interfering with state affairs, e.g., cyberter-rorism), denounce it as cybercrime (invoking laws against digital tres-passing, unauthorized accessing, tampering with data, or distributing viruses), or simply ignore it as having no strategic value. The best approach, therefore, is a balanced one that scrutinizes the actual online activities to determine the proper boundaries for the categorization.

When considering incidents of a cyberattack, an offensive act by nature, we can distinguish between cyberattacks by examining the identity of the perpetrators and the target, the method and frequency of occurrence, the goal to be attained, and the damage caused. The perpetrators can be individuals, loosely defined groups (e.g., a temporary association of people with a common goal), well-defined groups (e.g., established groups, coalitions, or organizations), or states. It seems that cyberwar should require state actors; however, online sabotage actions and cyberpartisanship may also be considered a paramilitary activity. The targeted networked computer may belong to a state government, a corporation, an affiliated organization, or other groups, or less probably to an individual. The frequency and duration may be limited to one occasion (single incident), expanded over some measurable length of time (campaign), or a mutually interactive series of offensive and defensive actions (engagement). Online attacks can aim to gain dominance by causing damage or compromising the opponent's information and communication system, to express disapproval, or simply to raise public awareness. In some cases they have no political motivation at all. The directly attributed damage of a successful attack can cause the disruption of system operation, loss of information, or just a bash on the political or public image of the opponent.

In relation to the context of politically motivated cyberattacks, the following categories for their occurrence are useful: in response to an incident or condition, as part of an existing conflict, as part of an ongoing militaristic campaign, and as part of an ongoing conventional war. The first two categories are apparently closer to the idea of hacktivism, while the last two could safely labeled as cyberwar. Since militaristic campaigns are usually carried out by states or (terrorist) groups we can consequently assume state involvement in the context of cyberwars, or at least a state-level agenda. Based on the analysis of cyberattacks above and their context, I will now attempt to differentiate between hacktivism and cyberwar.

In my view, offensive online actions fall into the following three general categories: cyberattack (isolated); cybercampaign (coordinated, part of an identified conflict); and cyberwar (sustained mutual engagement). The acts of hacktivism and cyberwar overlap in the second category. Hacktivism is a politically motivated single-incident online action, or a campaign thereof, taken by nonstate actors in retaliation to express disapproval or to call attention to an issue advocated by the activists. Hacktivism has been diversely labeled as online advocacy, virtual direct action, electronic civil disobedience, performance art, or cybercrime, cyberterrorism, and cyberwar, depending on the point of view. According to Tim Jordan, British sociologist of the hacker community, hacktivism is a social movement, a new type of direct action, an Internet-based activity centered on virtual politics (Conan 2001).

Hacktivists are either "wired activists," thus, activists adapting the Internet into their strategy, or "politicized hackers," meaning hackers per se now adopting political causes as the justification of their actions. Some activists worry that many of the politicized hackers are just regular hackers taking up the political cause du jour without truly believing in it in order to legitimize their activity or simply gain popularity.

When hacktivism is elevated to the state level (in agenda or in terms of actors) and when it becomes a sustained engagement between parties connected to an ongoing conventional armed conflict the goal of which is domination over the adversary, hacktivism becomes cyberwar. From a political activist's point of view, one goal of conducting a cyberwar is to subvert a political system. This special potential of cyber-militaristic strategies against authoritarian regimes has even been raised by the consultants of the government think tank RAND (Arquilla, Ronfeldt 1998, 46). Discussing cyberwars, Arquilla and Ronfeldt (1998) further noted that the revolution in military affairs (RMA) also means that nonstate actors armed with cyberweapons would play an increasing role in future conflict:

> The revolutionary forces of the future may consist increasingly of widespread multi-organizational networks that have no particular national identity, claim to arise from civil society, and include aggressive groups and individuals who are keenly adept at using advanced technology for communications, as well as munitions. (47)

Indeed, the advancement of information technologies, especially networked computer systems, and our reliance on them, coupled with the globalization of economies, politics, and culture, change the nature of information warfare. More technology is available; economies, everyday-life support networks, and militaries are increasingly dependent on computer networks. Social movements against globalization, or more aggressive resistances against worldwide corporate dominance, are ever more globally organized. Both on the physical and ideological level, the information and communication networks and the systems of knowledge of a society are increasingly vulnerable.

The United States is especially a potential target of global resistance given its grip on the economic and knowledge systems of other countries in the world. American foreign direct investment (FDI) controls many key companies of national economies. American media, through global broadcasting, transnational media corporations, and FDIs in local media companies, control many news and information sources around the globe. American consumer culture is infiltrating into the farthest corners of the planet, opening up closed societies and transforming their cultural habits to conform to a homogeneous U.S.-flavored one. For many self-conscious nations the Internet is the last battlefield upon which to resist American global dominance and cultural hegemony. Thus, we can expect more activism and even military conflict to occur in cyberspace.

Cyberprotests against the World Bank

In light of the terrorist attacks on the United States on September 11 the 2001 Annual Meetings of the World Bank (WB) and the International Monetary Fund (IMF) were canceled. Similarly, the wide-scale protests planned during the meetings have also been called off first, then converted into an anti-war rally. However, the horrific events of September 11—while having paradigm-shifting proportions in terms of national security and foreign policy—are not likely to change the agenda of the World Bank Group and the global social movement against corporate globalization.

Similar to other social and political movements that use the Internet (as discussed, e.g., by Ayres 2001; Cleaver 1998a, b; Cleaver 1999; Danitz and Strobel 1999; Denning 2000; Langman et al. 2000) the anti-WB protests used the Internet mainly for internal and external communication, education, and mobilization; sharing activist resources; and discussing logistical matters, such as transportation, accommodation, and provisions at protest sites. It would be difficult to think of any resistance groups that do not have their own Web presence. Web sites are the main source of information, especially for activists and sympathizers affiliated with these organizations. The recent anti-WB/IMF protests were orchestrated by such umbrella organizations as Anti-Capitalist Convergence (www.abolishthebank.org), 50 Years Is Enough (www.50years.org), and Mobilization for Global Justice (www.globalizethis.org). Furthermore, it has become customary for the coordinating entity to set up a specific Web site for each major protest, usually referring to the date of the action in the domain name (e.g., www.a16.org or www.s29.org). Yet another tactic of online advocacy is to create "spoof" Web sites with similar-sounding domains, criticizing and parodying the target entity (e.g., www.worldbunk.org or www.whirled bank.org).

Hacktivist actions, such as Web site defacements, virtual sit-ins, and e-mail campaigns, are also part of the protesters' strategies.The fascination with Web defacements may be explained by the multiversity of this act. On the one hand, it is cybergraffiti, a temporary disfiguration on the cyberfaçade of a company or organization. It usually leaves all other information intact, and even backs up the one modified index file. Such cybergraffiti is a brief critical intervention in the hegemonic status quo, "owning" or "rerouting" a symbolic gateway into the online establishment of a dominant power. This act of cyberprotest seeks public attention and visibility by delivering a political message through the media. Some see it as an online performance, an artistic expression of dissent. The "virtual sit-in," as its name suggests (with all its historic connotations) aims to block access to a service, in this case, usually a Web site. It is achieved by directing an overwhelming amount of coordinated data stream at the target server, which then radically slows down or crashes under the traffic. Depending on what ports are enabled on the server (MAIL, IRC, ICMP, etc.), these blockades can be achieved, for example, by flooding the server with a large number of e-mails that are beyond its capability to handle (e-mail bomb), overloading an IRC channel (IRC jamming), or overwhelming the server with small data packages that test the server's response capability (ping storm). There are, of course, many other, much more sophisticated attacks. Most of them work best if they are part of a coordinated attack, thus, executed simultaneously by a large number of machines (e.g., using the EDT's FloodNet Java applet for a ping attack, or in a distributed Denial of Service attack by pre-infected "zombie" machines).

On the other hand, Web defacements contain the potential for wreaking havoc, since hackers gain root access to the system, which means that they basically can do whatever they please. The chilling fact behind defacements is that they are carried out using the methods needed for launching a full-scale destructive cyberwar. In that sense, they do pose serious concerns. However, what keeps these concerns moderate is the fact that no sensitive or confidential information regarding national security or public safety is usually kept on public servers.

Activists have often come up with other innovative uses of the Net, such as the virtual march or Internet-transmitted laser-projected messaging to heads of states. During the Summit of the Americas meeting in Quebec City in April 2001, people who could not be physically present at the protests were invited to participate in a virtual walk on the www.marchedespeouples.org Web site by registering their names and nationalities. During the World Economic Forum (WEF) Meeting in Davos in January 2001 activists from all over the world could send a maximum of 160-character-long messages in English, German, French, Italian, or Spanish over the Internet either from the www.hellomrpresident.com Web site or over SMS messaging that were then projected onto a snow-covered hillside in 240 × 15 meter size with a laser beam. The projections could be checked real-time through a Web cam.

The spectacular and cutting-edge nature of these activities almost guarantees wide-scale media reporting. They involve high-tech global computer systems, brilliance and resourcefulness on the part of the hacker, illegal activity, secrecy, suspense, potentially enormous impact, and law-enforcement response. On the one hand, it has some folklorish elements, like a lone hero with a cheap computer and modem against an all-too-powerful government, a hostile foreign regime, or a monopolistic corporation. On the other hand, it may be framed as a story of criminals or cyberterrorists accessing highly classified military information, destroying vast amounts of essential data, or extorting money from international banks. Coupled with the public's general lack of understanding of the phenomenon, hacking has everything needed for a sensationalist headline story.

Hacktivists are certainly aware of the power of the mass media. As Denning (2000) notes, hacktivism contains the media as part of their strategy. Thus, if there is a message hacktivists wish to get across they will include it in the hack in some form, such as a note on a defaced or redirected Web site, in an e-mail bomb, or in a payload delivered by a virus. Alternatively, the disruptive act itself may call attention to the cause, mostly by the affiliation of the target site. This way, politically conscious hackers attempt to subvert the elite's control of the mainstream media by using it to their own advantage, hacking the process of reporting itself.

However, anti-globalization activists do not leave it entirely to the corporate media. In addition to the numerous protest Web sites and alternative press publications online, activists established their own globally organized voice by creating the Independent Media Center in Seattle to cover protests against the WTO in November 1999. It is now a collective of over fifty independent media outlets around the world, with hundreds of volunteer journalists offering grassroots, noncorporate coverage. Mainly through their Web site (www.indymedia.org), they are promising "radical, accurate, and passionate" reporting on issues—like major social movements—that are likely to be underrepresented or misrepresented in the mainstream media. More specifically, they offer up-to-the-minute reports, photos, audiostreams, and videostreams on their Web site, as well as via other independent media outlets. These reports and the corporate newsreels about the same event are often juxtaposed. During the anti-globalization protests, scores of media activists were equipped with digital cameras, distributing their footage over the Internet shortly after it was shot. Once the technology becomes more affordable and efficient, we can expect more instant video reporting, that is, real-time digital video streamed through a portable computer with wireless connection.

It would be a mistake to polarize the anti-globalization movement as a series of battles between the international financial institutions—such as the WB, IMF, WTO, WEF, or the G8—and the protesters. Rather, the global opposition movement unites a wide range of political voices, including anarchists, humanists, liberals, environmentalists, right-wing conservatives, and the radical left wing, against the political and economic practices of the developed West. Generally, they oppose the detrimental social and environmental consequences of the policies of the international financial institutions that benefit the rich at the expense of the poor in the name of deregulation, liberalization, and privatization. They target, specifically, the "structural adjustment" programs of the World Bank as pre-conditions of WB and IMF loans.

The anti-globalization movement has made its voice heard at major demonstrations in Geneva (WTO, 5/98), Birmingham (G8, 5/98), Köln (G8, 6/99), Seattle (WTO, 11/99), Davos (WEF, 1/00), Washington (WB/IMF, 4/00), Melbourne (WEF, 9/00), Prague (WB/IMF, 9/00), Quebec (FTAA, 4/01), and Genoa (G8, 7/01). As World Bank President James Wolfensohn (2001) pointed out, "protests have now become the norm, unfortunately, for virtually all major international meetings."

Responding to the ubiquitous protests at their major gatherings, the international organizations are now considering scheduling their meetings in remote, hardly accessible locations like the Canadian mountains in Alberta (G8) or the desert state of Qatar (WTO). They have also considered moving them entirely online (WB). Out of fear of demonstrations, for instance, the World Bank canceled its June meetings in Barcelona and decided to hold them on the Internet instead. The natural move for the protesters, of course, was to switch their street demonstrations to online action. As soon as the World Bank announced that their Barcelona meeting would be held online, protest organizers as well as the media reacted that it would, in fact, make it easier for Net-savvy protesters to disrupt the meetings by online activism tools, such as DoS attacks or e-mail spams (see Burke 2001; Ward 2001). However, the idea of online direct action splits the unified masses of activists since many of them are against breaking the law, and these actions indeed often fall on the illegal side.

Yet there are more factors that split the masses opposing globalization. For instance, Solomon (1999) describes the use of laptop computers by international nongovernmental organization *lobbyists* (as opposed to the more radical and technologically less prepared *abolitionists*) in their campaigning against the World Bank. She points out that laptops are a "double-edged sword for international activists. . . . While their usage of communication technologies is subversive of global power relations, it is also simultaneously reinforcing of hierarchical relations amongst their own networks" (78). What Solomon alludes to is a thought-provoking dilemma. If protesters start using laptops and other advanced communication technologies they basically subscribe to the Western corporate bias inherent in these technologies and will ultimately find themselves playing in the corporate league where they have a comparative disadvantage. Staying in the corporate ballpark, protesters and their strategies would allow the elite to frame them to fit into a popularly opposed category, such as crime, vandalism, or anarchy. The media would play along given their settlement for surface-level news and sensationalism. What the protesters have to seek instead is to disrupt the normalized operation of corporate hegemony in a way dissonant with established corporate practices.

However, despite Solomon's claim that the Internet is—like laptops—in Solomon's (1999, 78) words, "socially and normatively biased to favour hegemonic interests and exclude difference," I would like to think that the Internet remains different, especially given its still hovering open and free Barlowish origin myth. Even after the major commercialization and convergence in ownership and governance, the Association for Progressive Communications (APC) declared that the "Internet should be kept open and used to protect the environment, promote human rights, peace, development and democracy" (2001). Yet, this is not what is happening. As the corporate world continues to colonize cyberspace the inequalities of resources online are still fairly evident.

Apparently as an effort to provide a forum for dialogue and a source of information on global issues, the World Bank launched its Development Gateway Internet Initiative on July 23, 2001. While claimed by the Bank as a decentralized and independent forum where differing opinions are welcome, protest organizations quickly pointed out—allegedly based on a leaked document—that the true agenda behind this initiative for the Bank was to reach even more people with its perspectives and thereby countering the backlash against corporate globalization (Wilks 2001). The Bank hired Third Level Data, a marketing company, to promote the site as part of the $50 million budget and full-time staff of twenty-five behind the project. This is in sharp contrast with the limited resources of the protest organization, 50 Years Is Enough, the publisher of the *Economic Justice News*, which called for a "sustained anti-marketing campaign" by asking readers to send this article on to ten of their contacts and urge them to do the same. When asked about the online alternative for protests, should the World Bank decide to hold their meetings on the Internet, the president of the organization Njoki Njehu responded, "Are you kidding me? Just look around–I can hardly download my emails" (2001). Thus, a digital divide between corporate entities driving globalization and the protesters is obvious, despite the claims about the Internet as a great equalizer in activist movements.

The World Bank, on the other hand, seems very much prepared to defend its computer system against online attacks. The types of attacks occurring against the Bank include Web site defacements, DoS attacks, and viruses. Defacements succeed very rarely, according to a computer network security expert in the WB whom I interviewed on August 7, 2001. But even if they do there is a little chance it will be noted–he says–since the usual reaction time to Web site defacements is a matter of minutes, and has always remained under fifteen minutes (interview with the author in Washington, DC on August 7, 2001). The external sites are monitored constantly; they are compared to a mirror within the firewall to notice any changes in content. As a matter of fact, this security expert could not recall any defacements, except for one by a Hungarian hacker, but it was not anti-Bank motivated. He points out that the WTO, for instance, was not at all prepared for the Internet attacks that occurred during the Seattle meeting. Henceforth, their Web site was basically "trashed." He believes that the World Bank is much better off in terms of security. They are so concerned that they are willing to take the ultimate measures to protect their infrastructure, that is, to cut off their domain entirely from the Internet. For instance, during the Y2K scare they actually shut down incoming e-mail service for a period of time around midnight.

Internal rumors say that there are daily DoS attacks against Bank servers. The detection of such attacks, however, is almost impossible given the simultaneous distributed nature of attacks. The only deployed countermeasure was the shutdown of traffic from the identified IP addresses. The Bank claims it has not yet responded with counterstrikes. The number of virus attacks seems to have decreased over the last few years but this is mainly because of the multilevel virus protection in place (gateway level, server level, workstation level).

According to sources, the external public sites are the main targets of online attacks, as well as servers in the DMZ (demilitarized zone).[5] For instance, on June 25, 2001, hackers managed to break in and gained root access to the ibank6.worldbank.org server, stealing documents and compromising several hundred passwords. Allegedly, they were going to use the account of prominent senior managers to send out anti-globalization messages. While the Information Solutions Group (ISG), the WB unit responsible for network security, initially announced that "the ibank6 machine was possibly compromised by a hacker attack," later they retracted saying that "claims of break in could not be substantiated with hard facts" (e-mail from ISG staff on June 25, 2001). The hackers revealed their attack to the Spanish *El Mundo* and the Italian *La Repubblica* newspapers. Interestingly enough, this incident was not reported in the American media.

As revealed by internal sources, the financial losses as a direct result of Internet attacks converge to zero. However, the security of the external site is very important because of the image of the Bank and its security. The main purpose of these attacks is to embarrass the organization. As part of internal security measures, it is extremely difficult for anyone to obtain any data on the attacks against WB servers. People responsible for computer security would simply not reveal them: "The less you know about security, the better protected the Bank is." Furthermore, if security compromises are revealed it embarrasses not only the Bank, but also the information security system and its staff. Therefore, as this security consultant commented, "we are not even revealing everything to our management."

From the Bank's point of view, the hackers—however politically or ideologically motivated—are facing Internet security experts. These experts do not care about the political causes involved; their job is purely technical—to protect the computer system against attacks that seek to infiltrate or compromise it. The computer security unit is concerned about "the continued operation of the organization from a communication and information technology standpoint." They are even more concerned during "periods of high probability" such as the annual meetings.

Conclusion

As it has been alluded to in the text, the mass media constitute a very important battleground for activists. Media attention guarantees wide visibility and the most effective forum for publicizing the issues in contention. While the U.S. mainstream media are in the hands of the corporate world, the sensationalist nature of hacktivism works to the activists' advantage. Both the activists and the corporations are aware of the power of the image. Given the rich resources of corporate entities to fight hacktivism on the technological level, their main vulnerability lies in their public image. In cyberspace, company.com is the front entrance of their headquarters. Whether disrupted or defaced, it will be noted.

National governments are also represented symbolically by their Web presence, providing a tempting target for online action. Their physical and human resources are often lagging behind that of the corporate world, multiplying their vulnerability online. Yet the technology and know-how behind hacktivist attacks carry much more in possibilities than Web site defacements and virtual sit-ins.

Whether the targets of activism are the actors of corporate globalization or oppressive governments, the tendencies we see are very similar. On the one hand, the Internet is more and more integrated into resistance, substate actors are increasingly taking up state-level agendas, and their tools are nearly capable of causing damage of cyberwar proportions. On the other hand, governments and corporations are more and more prepared to deal with these threats, they are increasingly inclined to respond similarly with online counterstrikes, and we are just moments away from open government-government cybermilitary engagements.

While online activists perceive their missions as legitimate protests in cyberspace, the targeted corporations or governments regard them as online security breaches or national security threats, and seem to be willing to respond accordingly. As in the case of traditional resistance movements, by framing online activism as criminal activity or a national security threat, they reinforce their hegemonic grip on dissent.

Notes

1. See, for example, the 1995 murder of a labor activist in East Java allegedly by the military and the following Urgent Action (UA) posted on apakabar; the August 8, 1988, (8/8/88) uprising and its eleventh anniversary on September 9, 1999 (9/9/99), in Burma; and the June 4, 1989, massacre and its tenth anniversary on Tiananmen Square in China.

2. Burma was renamed Myanmar in 1998.

3. In a similar fashion, for example, the official White House Web site was parodied at www.whitehouse.org. Immediately following the tragic events of September 11, 2001, the Web site was taken down and offered condolences for the victim's relatives. In the case of the www.wtoseattle.org and www.seattlewto.org domains, however, both of them now point to a porn site (accessed September 30, 2001).

4. The idea of a virtual country gained popularity with those whose real countries were physically endangered. The virtual nation of Cyber Yugoslavia (CY) was launched on September 9, 1999. It is a nowhereland, occupying no physical space (yet) but offering CY citizenship and passports. They pledge, "When we have five million citizens, we plan to apply to the UN for member status. When this happens, we will ask 20 square meters of land anywhere on Earth to be our country. On this land, we'll keep our server" (www.juga.com). On September 30, 2001, the citizen count stood at 15,304.

5. DMZ servers are outside the firewall accessible to a predefined group of users.

References

Arquilla, J., and D. Ronfeldt. 1998. "Cyberwar is Coming!" pp. 24–50 in *InfoWar*, ed. G. Stocker and C. Schopf. Wien: Springer.

Association for Progressive Communications. 2001, February 22. PRAGUE "Leave the Internet Alone," Press release.

Ayres, J. M. 2001, August 29–September 2. "Transnational Activism in the Americas: The Internet and Mobilizing Against the FTAA." Paper presented at the Annual Meeting of the Political Science Association, San Francisco.

Burke, M. 2001. "Fearing Protests, World Bank Cancels Barcelona Meeting." Online. http://nyc.indymedia.org/front.php3?article_id=6008 (June 26).

Candeira, J. 2001. "Protestas Antiglobalizacion: Hasta La Cocina." Online. http://www.el-mundo-es/navegante/2001/06/25/esociedad/993454687.html (June 26).

Cleaver, H. 1998a. "The Zapatista Effect: The Internet and the Rise of an Alternative Political Fabric." Online. http://www.eco.utexas.edu/homepages/faculty/Cleaver/zapeffect.html (October 10, 2000).

——. 1998b. "The Zapatistas and the Electronic Fabric of Struggle," pp. 81–103 in *Zapatista! Reinventing Revolution in Mexico,* ed. J. Holloway and E. Pelaez. London: Pluto Press.

——. 1999. "Computer-Linked Social Movements and the Global Threat to Capitalism." Online. http://www.eco.utexas.edu/homepages/faculty/Cleaver/polnet.html (October 10, 2000).

Conan, N. (host), T. Jordan, C. Karasic, and O. Ruffin (guests). 2001, July 30. "Hactivism." *The Connection.* Radio broadcast. National Public Radio. WBUR, Boston. July 30.

Cult of the Dead Cow. 2001. "The Hacktivismo FAQ v1.0." Online. http://www.cultdeadcow.com (September 29).

Danitz, T., and W. P. Strobel. 1999. "The Internet's Impact on Activism: The Case of Burma." *Studies in Conflict and Terrorism* 22: 257–69.

Denning, D. E. 1999. *Information Warfare and Security.* Reading, MA: Addison-Wesley.

——. 2000, February 18. "Activism, Hacktivism, and Cyberterrorism: The Internet as a Tool for Influencing Foreign Policy." Paper presented at the Internet and International Systems: Information Technology and American Foreign Policy Decisionmaking Workshop, Georgetown University. Online. http://www.nautilus.org/info-policy/workshop/papers/denning.html.

Fielding, P., and D. Bennett. 2001. "Ten Steps to Building a Successful Online Campaign." Online. http://www.e-advocates.com/content/buildcampaign.pdf (August 15, 2001).

Havely, J. 2000. "When States Go to Cyber-War." Online. http://news6.thdo.bbc.co.uk/hi/english/sci/tech/newsid_642000/642867.stm (September 20, 2001).

Himanen, P. 2001. *The Hacker Ethic and the Spirit of the Information Age.* New York: Random House.

iDefense. 2001a, January 3. "Israeli-Palestinian Cyber Conflict (IPCC)." White Paper. Version 2.0PR. Fairfax, VA: iDefense Intelligence Operations.

——. 2001b, May 14. "U.S.-China Cyber Skirmish of April-May 2001." White Paper. Fairfax, VA: iDefense Intelligence Operations.

Krebs, V. 2001. The Impact of the Internet on Myanmar. Online. http://firstmonday.org/issues/issue6_5/krebs/index.html (August 15).

Langman, L., et al. 2000. "Globalization, Domination, and Cyberactivism." Presented at Association of Internet Researchers 1.0 Conference, University of Kansas.

Levy, S. 1984. *Hackers: Heroes of the Computer Revolution.* Garden City, NY: Anchor Press/Doubleday.

Miklaszewski, J., and R. Windrem. 1999. "Pentagon and Hackers in 'Cyberwar'." Online. http://zdnet.com.com/2100-11-513939.html?legacy=zdnn (March 5).

Njehu, N. 2001, August 21. Interview. Washington, DC: Voice of America.

OMB Watch. 1998. "Speaking Up in the Internet Age: Use and Value of Constituent E-mail and Congressional Web Sites." Washington, DC: Nonprofits' Policy and Technology Project.

Oppes, A. 2001. "Hacker Entrano In Azione Contro Forum Banca Mondiale." Online. http://www.repubblica.it/online/mondo/barellona/virtuale/virtuale.html (June 26).

Solomon, M. 1999. "Lap-Tops Against Communicative Democracy: International Non-Governmental Organisations and the World Bank," pp. 61–81 in *Technology and Public Participation*, ed. B. Martin. Wollongong, New South Wales, Australia: University of Wollongong.

Sterling, B. 1992. *The Hacker Crackdown: Law and Disorder on the Electronic Frontier*. New York: Bantam Books.

Taylor, P. 1999. *Hackers: Crime in the Digital Sublime.* London: Routledge.

Tedjabayu, Basuki. 1999. "Indonesia. The Net as a Weapon." *Cybersociology Magazine* 5. Online. http://www.socio.demon.co.uk/magazine/5/5indonesia.html

Ward, M. 2001. "World Bank Risks Attack From Net Protesters." Online. http://news.bbc.co.uk/low/english/sci/tech/newsid_1398000/1398485.stm (September 20).

Wilks, A. 2001. "Anti-Marketing Campaign Greets WB Internet Plan." *Economic Justice News* 4(2): 6.

Wolfensohn, J. D. 2001. September 4. Communiqué from the Office of the President to All Staff. Washington, DC: World Bank.

Wray, S. 1998. "Towards Bottom-Up Information Warfare: Theory and Practice: Version 1.0." Online. http://www.nyu.edu/projects/wray/BottomUp.html (February 23, 2001).

4

The Radicalization of Zeke Spier

How the Internet Contributes to Civic Engagement and New Forms of Social Capital

Larry Elin

The only things more consequential than getting arrested and jailed for what one believes are taking a serious beating or dying for it. All other forms of civic engagement seem tame, even trivial. Recall, for example, the face-to-face confrontations and hand-to-hand contests between political activists and the police that defined the street demonstrations of the '68 Democratic Convention in Chicago, during which demonstrators clashed with police in adrenaline-pumping, blood-spilling brawls. Hundreds were arrested, hospitalized, or both, and it was all on television. No matter where one stands on the issues, the passion, resolve, and courage of committed protesters make temperatures rise, and onlookers are left either emotionally inspired or intimidated.

The United States was formed by protesters: The First Amendment protects them, and folk ballads celebrate them. Protesting is not only an American tradition; the exercise of the right of protest is the civic equivalent of lifting weights. It strengthens us even as it underscores the issues that divide us. Researcher Karin Gwinn Wilkins (2000, 572) found that "Civic engagement and *political distrust* appear to be among the stronger predictors of political participation." People who have a squabble with the government are more likely to be civically engaged, to be politically active, and to vote. Mayors, police chiefs, and the local chamber of commerce tremble at the thought of a street demonstration by passionate contrarians while their city is in a convention spotlight. It's a public relations nightmare. But there is no stronger evidence of a healthy democracy than the presence of vocal dissenters. They are the physical embodiment of the First Amendment.

And so it was in Philadelphia, Pennsylvania, on August 1, 2000. Hundreds of demonstrators from all over the country shadowed thousands of Republican National Convention delegates as they descended on the city of brotherly love for a week of speeches, rallies, caucuses, and celebrations. The delegates settled into a regimen of carefully scripted activities scheduled well in advance by the Republican National Committee. The agenda was set. The delegates gathered by states in their assigned areas on the convention floor. Speaker after speaker hugged the microphone and methodically pounded down one campaign plank after another until the entire platform on which George W. Bush would run was soundly constructed.

Meanwhile, somewhere in Philadelphia, the demonstrators found each other, formed affinity groups, selected spokespersons, and reached agreement on the nature of the protests in which they would engage. Times, places, and topics were chosen. The level of civil disobedience was arrived at by consensus. Some protesters were willing to be arrested, if that's what it would take to make their point, and they went to one street corner. Others were not, and they went to another. And so it went, block by block, until they ringed the convention center with a virtual picket fence of anti-issues: corporate control, the death penalty, the criminal "injustice" system.

Inside the convention hall, the television networks and newspapers did their best to make an inevitable outcome, foreshadowed by events totally lacking in conflict or drama, interesting. Internet news sites covered the proceedings with high-tech streaming video from Web cams mounted on the ceiling or carried around by online reporters, believing, perhaps, that the technology really is the message.

Outside the convention hall, as far as two miles away, protesters were captured on video by police who later used the tapes as evidence in some of the trials of more than 350 people arrested that day. Included in the roundup by Philadelphia police was a nineteen-year-old veteran of vocal self-expression, group protest, and civil disobedience. Zeke Spier, a Brown University freshman who had already traveled to Georgia and Washington, DC, to march, was arrested with seventy-two others on the corner of Spruce and South Broad. Owing to the Internet at least partly, his is a story of awakening to social justice issues. Using the Internet, he prepared himself for political engagement, organized a demonstration, and assisted in his own defense in the criminal justice system.

Zeke Spier (his real name) was one of hundreds whom we interviewed for a book on the role of the Internet for the development of virtual political communities and new forms of social capital. *Click on Democracy: The Internet's Power to Turn Political Apathy into Action* (Davis, Elin, and Grant 2002) features profiles of more than twenty very different Americans who engaged in political action during the 2000 election largely through their intrepid use of the Internet. During our probing for interesting subjects for the book, we contacted the Philadelphia public defenders, who were themselves using the Internet to build a defense for those arrested during the Republican convention. They, in turn, put us in touch with their client, Spier. He agreed to an interview for the book, and we spoke to him by phone on several occasions—the first while he was still in danger of serving time in jail, and the last in 2002 after his case had been settled. Spier was an excellent addition to our book, as we looked for balance in representation across age, sex, ethnicity, political affiliation, and other demographics.

The Internet's information, communication, and networking power enabled Spier to discover his passion for social justice and to find the willingness to sacrifice his freedom to express it. Although other media stirred his emotions and informed his ideology, the Internet enabled him to conduct unmediated, two-way, one-to-one, and many-to-many communications with others of like mind. The Internet became for him the link between education and motivation and the catalyst for action. As we shall see, Spier's experience with the Internet is emblematic of how Esther Dyson, founder of Edventure Holdings and former chair of the Internet Corporation for Assigned Names and Numbers, sees the technology taking society to a level the traditional media cannot. The Internet, she believes, "is the tool of the non-establishment," and that will change the political power structure (Napoli 2000, 1). "The Internet is a medium of conspiracy, a medium of people not heard. It is profoundly disruptive. It asks you to talk back" (1).

Growing up in Portland, Oregon, Spier wasn't especially political, and his family, though regular voters, was not particularly active in civic affairs. He, his parents (both lawyers), and his two brothers rarely spoke about politics around the dinner table. But Spier kept up with current events by reading and watching the news. They always had the daily newspaper, the *Oregonian*, around the house and Spier developed the unusual habit (for a preteen) of getting up at 5 A.M. and watching the morning news shows on television. He speaks with confidence and intelligence about world affairs that belie his age.

In public high school, Spier was a solid student whose main extracurricular activity was theater. "My first experience of expressing political thought happened when I began working in school with people who wanted to do more topical theater pieces," he said. "I read books about the theater, particularly about Russian and German theater." European theater and politics are much more closely intertwined than in this country, and he was impressed with the power of theater as a means of political expression. He and his theater friends wound up writing and producing a play about the teenage immigrant experience. While doing the research for the play, they found themselves immersed in political issues because "a lot of what these people face as immigrants are because of government policies." The more he dug, the more he came to see current events in this country in a new light. "Some of the things about the way things work in this country made me angry."

High school classes in social studies and anthropology, which exposed Spier to historical injustices, furthered his immersion in political thought. He remembered a class that covered the United States' involvement in El Salvador. "This is stuff I was just outraged about," he said. "This is against everything I had been told about our country. Overthrowing a socialist in favor of a dictator." Eventually, this thinking led to exploring other spheres. He became a feminist by speaking with his female friends and finding out about their experiences. When he matriculated at Brown University in the fall of 1999, he was already cast as a socially conscious, politically aware reformer. By this point in Spier's life, the mass media—television, newspapers, and even theater—had played a major role in shaping his sense of self, but had contributed little to helping him find his place in a collective. "Who else is out there?" he wondered. He and his high school friends scattered, and lost touch with each other.

Spier sought out classes and activities that would buttress his ideas about social justice. He took a course called Civil Rights and the Legacy of the 1960s, but it wasn't academic work that drew Spier closer to the radical thoughts and actions that would eventually land him in jail; it was the personal connections he developed with like-minded students. Spier got on various listservs and e-mail lists at Brown that brought activists together. "One list led to another, and I started getting information about everything. Meeting updates, discussions, teach-ins, scheduled demonstrations, you name it." He started attending meetings and small, impromptu classes held by various professors, and these groups became part of Spier's extended community. The Internet became for Spier the doorway to a public sphere that was at times the classroom, the library, the dorm and even the Internet itself. Although the Internet did not replace face-to-face meetings and brainstorming sessions, it facilitated them, and in some cases enabled the discourse to extend temporally and geographically beyond the confines of the physical setting.

Spier the political activist was right at home on the Internet. Even though the Internet was developed by an agency of the Department of Defense, it was adopted early on by members of the counterculture who saw it as an excellent way to connect people with similar beliefs who were separated by time and space. Stewart Brand, who started The Whole Earth Catalog and was at the epicenter of several cultural revolutions in the San Francisco Bay Area, was also founder of The WELL in 1985. This was the cybercommunity Howard Rheingold glorifies in *The Virtual Community: Homesteading on the Electronic Frontier* (2000). On his Web site, Rheingold (1988, 1) describes its early inhabitants: "The Whole Earth network—the granola-eating utopians, the solar-power enthusiasts, serious ecologists and the space-station crowd, immortalists, Biospherians, environmentalists, social activists—was part of the core population from the beginning."

By October, Spier was ready for something beyond dialogue. Someone sent him an e-mail about the annual demonstration waged against the School of the Americas (50A), located at Fort Benning, Georgia. The U.S. Army runs the military school, and its students—military officers—come from Latin American countries. The left generally regards it as a training ground for assassins and military death squads that operate in Central America. Political groups across the spectrum suspect this may be true, and an amendment to close the school lost in the U.S. House of Representatives by only ten votes during debate over the Defense Authorization Bill for Fiscal 2001. The school's sixty thousand graduates include Manuel Noriega and Omar Torrijos of Panama, Leopoldo Galtieri and Roberto Viola of Argentina, Juan Velasco Alvarado of Peru, Guillermo Rodriguez of Ecuador, and Hugo Banzer Suarez of Bolivia. Some of its graduates are tied to the most gruesome massacres of civilians in El Salvador, Guatemala, and Honduras. When Spier heard of the demonstrations to shut the school down, he said, "That's for me. I'm going." He boarded a bus and, along with a couple hundred other students from Brown, joined ten thousand protesters who "crossed the line" onto federal property at the School of the Americas, risking arrest.

Spier wasn't arrested, but saw what he called "the police resistance to protesting." "The police lied to us, they misled us, they did things to break us up," he said. He described how, at one point during a solemn march to mourn those killed in Central America by SOA alumni, the police unleashed an ambulance through the crowd, sending the protesters scurrying off the road. "It was a ruse. It just stopped and didn't pick up anybody," he sighed. When he first went down to Georgia, Spier thought he'd get a chance to have his voice heard, he'd be empowered, and then he'd go home. However, the behavior of the police toward the protesters alarmed and awakened Spier to the cost of exercising free speech. "It radicalized me," he said. About a month later, Spier heard about demonstrations against the World Trade Organization (WTO) in Seattle, and the violent reaction of the police strengthened his resolve to protest again. "The adrenaline," he said, "was flowing."

"The first place I went to was the computer to find out about the WTO," he said. "I heard about the demonstrations in the mainstream media, and I did not think it was the full picture of what happened. I didn't trust them." The traditional media, after all, did a poor job of reporting what he had just experienced in Georgia. "I went on a bunch of searches and found the Independent Media Center Web site (indymedia.org) in Seattle," a site he would rely on again and again as his thirst for participating in civil disobedience grew. He spent hours and hours on the site, looking at pictures and video, reading firsthand accounts, and constructing his own image of what had occurred in Seattle.

Many scholars are concerned that the Internet is an unreliable source of political information, and, when used as a communications tool, they express concerns about its stultifying effect on public discourse. Rita Kirk Whillock (2000, 72) argues that "political uses of the Web have deprived individuals of an effective public voice while perpetuating a voice that is of more value to the propagandist than to the group of individuals involved." Cass Sunstein (2001), in his book *Republic.com*, maintains that the Internet is replacing the physical public space where citizens are exposed to different points of view with a private place where individuals withdraw into themselves and reinforce deeply held prejudices. And Robert Putnam (2000, 171), who traced the steady decline of social capital (in which political information and discourse play an important formative and maintenance role) in his book *Bowling Alone*, laments, "The absence of any correlation between Internet usage and civic engagement could mean that the Internet attracts reclusive nerds and energizes them, but it could also mean that the Net disproportionately attracts civic dynamos and sedates them."

Nevertheless, Spier, neither nerd nor dynamo, surfed the Web looking for other sources of information about the WTO and its activities throughout the world. Among the sites he visited were sites put up by the WTO itself and its members. He concluded that it is anti-democratic and anti-poor. He could not be in Seattle, and he needed an outlet for his feelings, so he did an art project about the subject for one of his classes at Brown, but that was only a panacea. Shortly afterward, indymedia.org announced a demonstration against the International

Monetary Fund (IMF) and the World Bank in Washington, DC, in April 2000. Since it was an easy drive from Providence, Spier could hardly wait to go.

Spier used the Internet extensively to organize the trip to Washington among his Brown peers. "We advertised our organizational meetings on the independent media Web sites and on all the e-mail lists." These were often unmoderated lists. Somebody would set them up, people would register, and then they'd take on a life of their own. Spier co-opted the sites for a time, monopolizing the discussions to build momentum for the action. He also found it necessary to "get smart" about the IMF, so he researched the organization. "About 95% of the research I did was on the Internet. I visited the IMF official Web site, then I would go and do a search to find all the protest groups—20 or 30 groups. I wanted to make sure I was informed enough, so I went after the facts, myself. It's more than a passion, it's an intellectual exercise." When April rolled around, Spier traveled to DC with about one hundred other students by bus.

In what Spier described as "a halfway futile attempt to shut the IMF meeting down," he and his band of demonstrators were assigned an intersection to blockade. Other groups did the same to establish a perimeter around the building. "We woke up at 4 am one day, went to the location, and sat down in the intersection at 18th Street and I. Some of us handed out information, others chanted," he recalled. The police did not bother his cadre, but Spier saw others who had been hit with pepper spray stumble by. "That helped radicalize me, further," he said.

For the first time he caught a glimpse of a group called the Anarchist Black Bloc, a tradition, he says, that came from Germany, England, and Italy. These protesters dress identically in black, both as a statement of their political beliefs and also as a tactical effort to avoid being singled out by the police. "People in Europe are more militant than they are here," said Spier, "and it's pretty common in Europe for somebody to turn over a car or set something on fire." The black-clad militants can melt into the similarly dressed crowd, and avoid detection. The Black Bloc idea was used by some WTO protesters in Seattle, and then again in Washington, DC during the IMF protest. "The American version (of the Black Bloc) didn't destroy property. They acted as decoys to attract the police and lure them away from the peaceful protesters. I actually saw them position themselves between a line of demonstrators and the police who were approaching. I was really impressed by that. They were the reason I didn't get pepper sprayed."

According to Spier, the Black Bloc is so loosely organized that the participants do not even know each other. They simply share a radical, anti-capitalist view. They find each other, and plan their civil disobedience, entirely on the Internet. Infoshop.org, which bills itself as "your online anarchist community," is one Web site that features news and information for Black Bloc activists. It provides the latest headlines for demonstrators looking for the next chance to protest against everything from capitalism to racism. "Of course, they keep things very vague because they don't want the police showing up where they show up," he laughed.

The presence of left-wing political groups such as the Black Bloc on the Internet has scholars such as Andrew Calcutt (1999, 2) debating whether the Internet will be the end of the nation-state, because of the way it empowers these individuals, or the advent of Big Brother, because of the control that could potentially be exercised by a reactionary government. On balance, the Internet community continues to behave in an anarchistic, somewhat anti-authoritarian manner, even as its numbers swell with average, everyday people. What was designed as a technological marvel—one that can avoid down-time by routing itself around nuclear blasts—has become a cultural marvel in the way that it routes itself around corporate control, censorship, and the authorities.

Even the *soul* of the Internet would seem to be fit for the counterculture. David Legard (2000, 1), writing for the Singapore Bureau of the IDG news service, analyzes the Internet from the point of view of Eastern philosophy, ruminating on whether or not the Internet is a yin or a yang construct. "According to the ancient I Ching and Tao Te Ching texts, yin represents all that is informal, consensual, and flexible and yang represents all that is orthodox, formal, and rigid. . . . The Internet—anarchic, personal and anonymous—is pure yin," he said. "It is a perfect fit with modern Western notions of individual freedom superseding communal benefit, democracy before prescribed stability."

The Black Bloc, like many other anarchist and protest groups, has no actual "membership." Spier explained: "They are simply a group of people who decide to use the same particular tactics during one particular action," he said. "It's not something you're in or not in. It's just something you choose to do, or not do. And it wouldn't be possible at all except the Internet gives all of these people a chance to organize, and not get caught." Whether they are, in fact, being chased or even observed closely by the police or other authorities no one can say for sure. But anarchist groups are certain that their activities threaten powerful interests, and that those interests control the law and order systems. And as an adage goes, "Just because you're paranoid doesn't mean you aren't being followed."

To Spier, more important than not getting caught is actually making a difference. By the time he was sitting on the pavement at 18th and I, he had traveled about one thousand miles to two different protests, had spent hours on the Internet indoctrinating himself, and had spent days away from school in strange cities. That morning, he had gotten up earlier than any self-respecting college student ever does and had barely escaped getting a bronchial condition courtesy of the Metro police. Did it matter? "I went into a sandwich shop just to take a break, and sitting there was this businessman. We started talking, and it turned out he was a delegate to the IMF conference from Europe. He couldn't get in because we were blocking the way," Spier said, with a bit of surprise in his voice. "We actually got into a decent dialogue about the protests and about the IMF. It was a marvelous exchange. I actually think he learned more from me than I did from him. He hadn't thought about a lot of the concerns that people were raising. I didn't change his mind about anything, and he didn't change mine, but it was a terrific discussion."

Four days later, exhausted and tired, "We slept on the floor of a Jewish Community Center, but actually got very little rest." Spier took a bus back to Providence. He finished his freshman year uneventfully, bought a car in June, and started back home to Portland by himself. "I had already heard, over the Internet, on the listservs I was on and on all the sites I usually go to, that there were protests planned for both the Republican and Democratic conventions," he said. He planned his road trip so he could be back east, in Philadelphia, by the end of July.

For Spier, the Republican convention offered an opportunity to protest a number of the party's policies with which he takes issue. His big three were the degree of power that corporations exercise over the party, George W. Bush's record with the death penalty, and the drug war, which Spier described as "a war on the poor, and in particular Black people." "This country just went over 2,000,000 people in jail. That's more than Iran and China. All of these issues, and the proof behind them, are on the Web site that you can go to, to learn about the protest. It's all there, as well as links to many other Web sites," he explained. "I don't automatically believe that everything on these sites is true, but there are levels of cross-referencing that I go through to get at my final sense of the truth. There are levels of trust, and I trust the ACLU, for example, and Amnesty International." Spier was convinced he was on solid moral, ethical, and legal ground when he decided to exercise his free-speech right in Philadelphia later that summer.

During his road trip to and from Portland, his only contact with the Philadelphia protest organizers was by Internet. "I stopped at public libraries and checked the updates on the convention, how the organizing was going, any updates on the police, whether permits had been gotten for any of the marches, and so on. It ended up that I was able to get there [to Philadelphia] on July 27th, and my first stop was the public library. I got on the Internet and looked up where the convergence locations were." Convergence locations are places where teach-ins, training, and gatherings take place when protesters come in from out of town. "If it weren't for the Internet, I don't know how I would have found anybody," he said.

What he found, once he got oriented, was an affinity group that fit his particular tastes—a group whose protest goals matched his own. In protest parlance, this is a group of from five to twenty-five people "who establish a consensus-based structure of like-minded individuals to participate in an action," said Spier. He explained that in affinity groups, which trace their origins to the Spanish anarchist movements of the 1930s, "People have different roles—a support person, a media person, somebody who can interact with the police." Eventually, each group chooses somebody to represent them at a "spokes council," where the entire protest is planned. In this way, directives come from

Who we are

The R2K Network is an umbrella group of organizers, coordinators and activists planning to demonstrate, educate and agitate before and during the upcoming Republican National Convention. The network includes Ad Hoc Committee to Defend Healthcare, Unity 2000, the Kensington Welfare Rights Union, the Philadelphia Direct Action Group, and other groups planning or participating in Convention-related actions.

Some R2K members are engaged in direct action and civil disobedience; others have "worked within the system," though with some difficulty, and obtained permits and city services for their events. All are committed to peaceful, non-violent political activity.

Why we are

The political process is morally bankrupt. Both the Republican and Democratic parties neglect the interests of poor and working people and cater to the large business interests who fund their campaigns. These interests have used the parties to block environmental initiatives and roll back people's welfare programs while increasing corporate welfare programs. Average citizens who have suffered the consequences of these changes realize that they have less access than ever to the political process. Many people have responded by choosing not to vote. By protesting the convention, the R2K Network is sending a message that those who reject the two party farce are neither apathetic nor passive. Protesting this convention is a demand for political accountability, radical democratic action and an end to policies that hurt people and the environment.

The protests in Philadelphia have been enthusiastically joined by thousands who intend to address the hearts and minds of the national populace with these problems while politicians rub elbows with the corporate lobbyists who ply them with lavish dinners and fundraising events behind closed doors.

Reprinted from the R2Kphilly Web site used by Spier to keep up on the protest plans for the Republican convention, http://r2kphilly.org/index02.html (January 18, 2001).

the bottom up, instead of from the top down. "The idea is to arrive at a level of involvement that everybody feels comfortable with, but to maintain a level of independence," Spier explained. "And another reason is to avoid being infiltrated by the police. At the affinity-group level, you know everybody."

While the Internet acted as the networking tool to bring these activists together from across the country, the face-to-face meetings and the formations of affinity groups were the essential step in forming the community that would act as one on the street. Strong arguments are made by those who believe there can be no "out of body" community, that it could not have formed exclusively on the Internet. It is impossible, they say, for people to form communities if they do not or cannot interact face-to-face in shared public space. There must be actions between them involving discourse, contracts, promises, and shoulder-to-shoulder heavy lifting. These are not simply symbolic acts, but real acts of conjoined efforts that result in tangible accomplishments.

The social theorist Ferdinand Tönnies ([1887] 1953) used the term *gemeinschaft* to describe a type of community in which people share "real and organic life," and common beliefs, needs, goals, and rewards for living and working together. While Tönnies lived and wrote during a time when much of Europe swelled with small towns and cities where the *gemeinschaft* ideal was self-evident, in today's America it is difficult to find in large cities and suburban areas. Indeed, Putnam (2000) chronicles the decline of all forms of social capital that the *gemeinschaft* embodied. And yet, in many ways, Spier's impromptu affinity group—which owed its existence at least partly to the networking on the Internet—was a modern day *gemeinschaft*, albeit a short-lived one.

"Without getting into fuzzy legal areas, what with my trial pending, people in my group took part in a protest at South Broad Street in Philadelphia. About 100 to 150 people protested there, and that's where I was arrested." Spier had co-organized the protest, which focused on the damage caused to young people because of some policies of the Republican Party. One issue he singled out was the party's support of mandatory sentencing in the criminal justice system and the adverse affect it has on the children of men and women incarcerated for minor offenses, many of whom are people of color. Most of the protesters were, themselves, young. Spier estimated the average age was between seventeen and twenty. He was quick to point out that they were not supporters of Al Gore, and in fact intended to protest in exactly the same manner at the Democratic Convention the following week. He wouldn't make it.

When the police arrested Spier and about seventy of his fellow protesters they were shackled, hands behind their backs with plastic tie-straps, and loaded into busses. "I was charged with four misdemeanors," he lamented. "Usually, somebody arrested for protesting would be charged with a summary offense, spend a night in jail, and that's it. That didn't happen here. Each charge against me carried a five year prison term." Spier and nearly all of the others arrested at Spruce and South Broad that day were charged with obstructing a highway, disorderly conduct, resisting arrest, and conspiracy to commit the other three misdemeanors. During his arraignment, Spier exercised yet another form of protest—perfectly legal and used by many in mass-arrest situations like this one—of not giving his name. The practice is known as "jail solidarity," and it usually causes havoc for the arresting party, who winds up with a jail full of John and Jane Does. For the strategy to be effective, everyone arrested must agree to refuse to provide their name, something that requires that everyone trust each other. Spier became a John Doe, and joined a number of others who refused to identify themselves to the court. He was unable to come up with $10,000 bail.

Little by little, others capitulated and gave their names to the police. Eventually, the solidarity eroded and, after nine days in jail, Spier gave in. He gave his name to the police and was released on his own recognizance. The Democratic Convention, which had been his next target, happened without him. Many of those arrested with him worked things out on their own, or went to trial like he did on September 23. At Spier's trial, he was one of forty-three codefendants who were brought up on charges. The district attorney presented a case against only five of them, including Spier, while charges against everybody else were dismissed. The most damning evidence against him was a video shot by the police that showed Spier sitting in the street. He was found guilty on two counts, obstructing traffic and conspiracy to obstruct traffic, and was released on appeal.

Spier never went back to school, and instead took a room in a house with several other young activists in Philadelphia, "where it's really not expensive to live," he said. He signed on with R2Klegal, an organization set up before the convention to support and coordinate the legal battles of anybody arrested during the convention. Anticipating that there would be at least some arrests during the protests, R2Klegal had arranged for some private attorneys to work on behalf of the protesters pro bono. However, more that 420 (by some accounts) were arrested, far more than the private lawyers were willing to defend for free, and adding significantly to the coordination of the legal struggle.

The private attorneys were joined by public defenders, and the two groups worked closely to share information under the umbrella of R2Klegal. Spier did legal support for those arrested, helping them in court, dealing with the lawyers, holding press conferences, and doing other media work. A significant part of the defense effort was handled over the Internet, using the organization's Web site, R2Kphilly.org, as the base of operations. "We also have a listserv of the defendants, and we keep everybody informed about court dates and other updates," said Spier. "The Internet is the main way we communicate with everybody involved." Although he is not the Web master (the young man who is was also arrested and found not guilty), Spier posts messages on the site.

About fifty-five defendants still had trials pending on January 20, 2001, even as George W. Bush's inauguration occurred two hours south on I-95. The majority were college students who hailed from thirty-five states, and keeping them informed about rescheduled court dates, new evidence, new witnesses, and other matters was a daunting enterprise for a grassroots operation like R2Klegal. The Internet—cheap, fast, and dependable—provided the group with its only viable communications and networking tool.

Sean Nolan and Meg Flores, two public defenders, were assigned to about three hundred cases stemming from the demonstrations. We contacted them and interviewed them by phone for our book after reading about their efforts on the R2Kphilly Web site. Typical of public defenders, thirty-seven-year-old Nolan and twenty-nine-year-old Flores are hard working, idealistic, and committed. Both come from the suburbs of Philadelphia, where they were raised by politically active, liberal parents. They share a passion for defending the poor, and they share many of the political beliefs of their clients. Perhaps because of that, and their professionalism, they have gone the extra mile in their defense of the convention protesters.

Nolan and Flores were accustomed to handling individual cases, not a large number of clients whose arrests occurred on the same day in various places throughout the city. They had to find a new way of coping with the logistics presented by the huge caseload. They adopted the Internet to conduct their own investigation and develop a defense strategy. "There are still several felony cases we are handling. These people were charged with assaulting police," explained Nolan. "We think it happened the other way—they were themselves assaulted."

To gather evidence, Nolan and Flores visited the indymedia.org Web site where photographers had posted pictures of the demonstrations at the convention, some of which seemed to contradict what the police account. They contacted the photographers by e-mail and got additional pictures and statements from the photographers who had witnessed the arrests. The network of contacts grew from there until the attorneys had a file thick with affidavits and eyewitness accounts.

"Trial dates get changed, we file tons of motions, and we've been able to use the Internet to keep in touch with our clients. We just do mass e-mails," explained Flores. Most of the clients Nolan and Flores usually represent are indigent—they don't have e-mail addresses, or real ones, either—but the convention arrests were primarily students, and they all do. Not only were Nolan and Flores able to keep the defendants informed, but, as Nolan described it, "It's been helpful to them as well, because they are all trying to act in solidarity with each other, and they have been able to keep in contact with each other." "They are an extremely committed group of individuals," added Flores.

Zeke Spier is more than committed. He became, as he willingly admits, incorrigible. While awaiting retrial in Philadelphia, he heard about another protest (again, over the Internet) in November 2000, in

Cincinnati, Ohio, against the Transatlantic Business Dialogue, a conference for the top two hundred CEOs from the United States and Europe. "This time I was arrested completely illegally," he says. "I was arrested for chanting."

By the spring of 2002, Spier's legal woes were over. "Minutes before my jury trial in Philadelphia the DA offered ARD, or Alternative Rehabilitative Disposition with no fine. This means they would drop the charges in 6 months if I was not arrested again," Spier explained in an e-mail. "After the district attorney presented her case in Cincinnati, which consisted of one officer testifying that I might have been obstructing pedestrians, the judge granted our motion to dismiss. I now have a civil rights lawsuit pending."

Spier is living in Bellingham, Washington. He plans to attend Whatcom Community College in fall 2002, and after that Western Washington University. He will major in education.

Spier's use of the Internet to become informed, to communicate, and to organize for activities that he then participated in physically seems to be emerging as a formula for civic engagement among a broad range of Americans. We encountered this pattern over and over during the research and interview phase of our book among gun-rights, gay-rights, and voting-rights groups and many others. In case after case, individuals with similar beliefs or identities located and networked with each other from coast to coast, if not face to face. Once formed on the Internet, these virtual political communities often moved into the physical sphere, as Spier did, to engage in political action. In their dealings with each other over the Internet they displayed a high degree of trust, honesty, and reciprocity, the coins of the realm of social capital and absolutely necessary for the creation and maintenance of community. Quite the contrary to Putnam's earlier quoted assertion that "the Internet attracts reclusive nerds and energizes them . . . [or] attracts civic dynamos and sedates them" (Putnam 2000, 171), Spier and many others are civic dynamos who are energized by the Internet. One could conclude that the term cyberactivism, the title of this book, describes not only an activity that takes place in virtual space, but also the chronological *process* or *path* that leads activists from the information highway to the streets.

References

Calcutt, Andrew. 1999. *White Noise: An A–Z of the Contradictions in Cyberculture.* New York: St. Martin's Press.

Davis, Steve, Larry Elin, and Grant Reeher. 2002. *Click on Democracy: The Internet's Power to Turn Political Apathy into Action*. Boulder, CO: Westview Press.

Legard, David. 2001. "Yin, Yang, and the Internet." Online. http://www.idg.net/go.cgi?id=399017 (January 19).

Napoli, Lisa. 2000. "Empowering Voters via the Net." Online. http://www.msnbc.com/news/445947.asp (April 14).

Putnam, Robert D. 2000. *Bowling Alone: The Collapse and Revival of American Community.* New York: Simon and Schuster.

Rheingold, Howard. 1988. "A Slice of Life in My Virtual Community." http://mirrors.ccs.neu.edu/EFF_Net_Guide/eeg_263.html (July 20, 2001).

——. 2000. *The Virtual Community: Homesteading on the Electronic Frontier.* Cambridge, MA: The MIT Press.

Sunstein, Cass. 2001. *Republic.com.* Princeton, NJ: Princeton University Press.

Tönnies, Ferdinand [1887] 1953. *Gemeinschaft und Gesellschaft* [Community and Society]. Trans. Charles P. Loomis. New York: Harper and Row.

Whillock, Rita Kirk. 2000. "Age of Reason," p. 172 in *Understanding the Web: Social, Political and Economic Dimensions of the Internet,* ed. Alan B. Albarran and David Goff. Iowa: Iowa State University Press.

Wilkins, Karin Gwinn. 2000, Fall. "The Role of Media in Public Disengagement from Political Life." *Journal of Broadcasting and Electronic Media*: 569–80.

PART II

Theorizing Online Activism

5

Democracy, New Social Movements, and the Internet

A Habermasian Analysis

Lee Salter

Introduction and Clarification

It is perhaps unsurprising that the theoretical constructs of one of the most important social philosophers of the present time have been applied to the study of the Internet. These theoretical constructs offer not only a clear factual explanation of how democracy has become subverted, but also, more recently, a normative guide out of the impasse envisioned by his predecessors. Habermas (1996) has recently attempted to apply his vast theoretical framework to a proposal for a constitutional democracy. In so doing, he has attempted to provide an explanation of how flows of influence may be organized so as to allow the most extensive democratization as possible, without that democracy becoming subverted by systemic imperatives. It has seemed clear for some time now that the informal layers of political society identified by Habermas have suffered a communicative deficit that may well be filled by a medium such as the Internet. In various discussions, the work of Habermas has been used as a theoretical backdrop to the claim that either the Internet provides citizens with a public sphere, or that it does not.[1] To be sure, there have been numerous conflicting accounts of the degree to which the Internet does or does not constitute a public sphere. Many such discussions suffer one or both of two fundamental flaws: Either that they fail to make use of an appropriately hermeneutic methodology, such as that employed by Dahlgren;[2] or they have tended to restrict themselves to the notion of the public sphere developed in *The Structural Transformation of the Public Sphere* (*Structural Transformation*), mentioning other aspects of Habermas's work only in passing, if at all. However, to be sure, there is a wealth of insight and analysis in the work of Habermas; much of this has yet to be employed in analyses of the Internet.

In the first section of this chapter, I provide an overview of the contemporary discussion of the public sphere, adding few substantial contributions. I then briefly introduce the main implications of the "system/lifeworld" dichotomy developed in *The Theory of Communicative Action* and the more sophisticated conceptualization of the public sphere that Habermas advanced in *Communicative Action* and *Between Facts and Norms*. Next I argue that the Internet is well positioned to facilitate communications in the less formal sense layer of political society, introduced in *Between Facts*. Finally, I illustrate how social movements shape the Internet to suit the form of communication appropriate to their interests, and how successful they are in so doing. To this effect, I examine one of the most important facilitating movement groups, the Association for Progressive Communications, and its relation to the Internet. Before going any further, it is necessary, and good practice, to clarify the object of analysis. To that end, I want to explain what I mean by the "Internet," and I hope to pre-empt concerns over the question of technological causation.

The Internet has become an unclear concept. Whereas it is recognized in dictionaries as a noun, it is also subject to normative description and use, which, to paraphrase Wittgenstein, determines its meaning. Indeed, when Nortel Networks ask the reader of their advertisements what they want the Internet to be, they illustrate just this point. The Internet is, *at base*, merely a network of computer networks. However, who can use it, for what purposes, and with what restrictions or support are all questions that advertisers, companies, governments, and civil society organizations are trying to answer at the moment.

Indeed, one pertinent question asked by Calabrese and Borchert (1996) is whether we might be able to regard the Internet as a social good and thus encourage its provision under the auspices of public service broadcasting. So, when referring to the Internet, I am not restricting myself only to the technical definition of a computer network, but rather to the social construction of it, and I will return to this question in due course. It is worth making a distinction between different Internet technologies in this context. The Internet is accessed by various applications, such as Telnet, bulletin board systems (BBS), Usenet, e-mail, and the World Wide Web.

The Web was initially just one part of the Internet, linking text documents together via hyperlinks. Hypertext mark-up language and the hypertext transfer protocol used for the www has evolved so that it can now be used to access most other Internet applications. Thus the Web has not only made Internet access available to millions more people than any other Internet application, but it has also made the Internet easier to use in a technical sense, and has reduced the number of separate applications required in order to access it. Further to this, the Web is now the main access point to the Internet, the application that people are most familiar with.[3]

An additional cautionary note must precede the following discussion. Much of the most insightful and original writing about the Internet seemed to have taken place in the mid 1990s, prior to the expansion of the Web. Thus, essays such as Poster's (1997) were essentially addressing BBS, e-mail, and the array of local area networks, such as the Santa Monica Public Electronic Network surveyed by William Dutton (1996). There is, then, a methodological difficulty in comparing, for instance, Poster's comments on the implications of the Internet in 1995 with the implications of the Web in 2002. One such difference is that between the 30 million Internet users that Poster wrote of, a large proportion of whom were computer enthusiasts, and the 300 million users of the Web today. Such difference consists not merely in a quantitative sense, but also in the notion of qualitative difference between constituencies. In addition to this, as earlier, the Web connects up other Internet applications so that it is more difficult to make distinctions between them. Nevertheless, it is still important to recognize which application is the object of analysis; that is, an analysis must note whether and when it is addressing the Internet as a whole, or a particular application.

Further to this is the question of the direction of causality between the Internet and society. Various writers, such as Martin Heidegger, Alvin Toffler, Marshall McLuhan, and Mark Poster have found themselves accused of technological determinism in relation to information technologies. Even for Habermas (1971) "there is an immanent connection between the technology known to us and the structure of purposive-rational action" (104–5) such that the former necessarily caters to the latter interests. I shall make only a few brief comments on this matter, which I hope will sufficiently depict my position. The notion that a new technology has some *necessary* impact upon society is wrong. "Strategies" of interpretation and implementation mean that the impact of a particular technology on a range of distinct societies is often diverse. The society, and groups within it, will interact with the technology both before and after the design process, shaping it and modifying it to suit their own practical prioritized objectives. Indeed, different groups and classes in a single society will have divergent (often conflicting) interest requirements of a technology and will struggle to control its implementation in accord with these. To adequately assess a technology, it is important to realize the range of interests, and those that attempt to dominate discourse, while recognizing that struggles take place between interests. So as to avoid, on the other hand, excessive social determinism (see, for example, Winston 1998) whereby technological rational of domination "is the result not of a law of movement in technology as such but of its function in today's economy" (Adorno and Horkheimer 1997, 121), the degree to which a technology is "closed" to interests, in the short and long term, must also be considered. Such an approach allows us to consider, in addition to a plurality of interests effecting the development of a technology, that the influence of these interests on a technology *changes* over time. Technological development is an ongoing process.

Notwithstanding this, to say that a particular technology has no intrinsic qualities is equally inaccurate: No matter how much social shaping takes place, it is absurd to suggest that a television can be used to wash clothes. Thus, a cautious balance must be held between the transformative capacities of a technology on the one hand, and the capacity of social agents to utilize technologies, and shape them in their use, on the other hand. This interplay will be held to be of utmost importance in this chapter.

The Public Sphere, Lifeworld, and Colonization

Structural Transformation has been rightly criticized for a number of historical and theoretical omissions and suppositions (see Calhoun 1992). Although Habermas has accepted many criticisms, such as those pointing to his exclusion of the idea of working-class public rationality, he passed over the opportunity to make changes to the 1989 English translation, believing the general points to still hold validity. Indeed, it would be folly to deny the importance of the book, especially in view of its important critique of modern capitalist democracies.

Habermas's central thesis in this work is that during a period of epochal change, there arose independent forums for rational-critical debate. In the spirit of the bourgeois revolts, the relationship between title, status, and voice was eliminated in the public sphere of the eighteenth and nineteenth centuries: It was *formally* open to all irrespective of class. In the bourgeois public sphere, arguments stood or fell in accord with the power of the better argument rather than with the power of coercion. However, once the bourgeoisie had consolidated their hegemonic position, their public sphere, which employed, or was founded upon, the public use of reason to critically challenge authority, became an empty concept. The entry of diverse claims of the working classes, of women into the public sphere, and the influence of heterogeneous cultural products conspired to drive out the universalistic ethos of the homogenous public sphere. During this process of "refeudalization," of the disintegration of dialogical publicness and its replacement with public displays designed to gain popular consent, what was once referred to as "public opinion" became something to be manipulated and at best consulted rather than something autonomously generated by rationally debating citizens.

I do not intend here to get involved in the debate over the accuracy of Habermas's historical account. Nevertheless, it must be accepted that as a normative critique of capitalist democracies, there are important lessons to be learned, not least that a democratic government is one that acts upon the *genuine* will of the people, that is, a general public will rather than isolated particularistic interests. A number of writers who specialize in the study of the Internet have taken the account of the bourgeois public sphere more or less as given, and have tried to project it onto the Internet. The folly of such an exercise consists not only in that the bourgeois public sphere arose in a specific period of legitimation crisis, but also due to the supposedly commensurable objectives and interests of participants in the bourgeois public sphere, against the plurality of Internet users. In fact, referring solely to *Structural Transformation*, it seems that the only commonality is that they are both formally open to all (notwithstanding the fact that in practice "all" in the bourgeois public sphere meant all aristocratic and bourgeois males). To be sure, the fact that the bourgeois public sphere sought to form a common will, whereas that Internet seems to fragment or at least question the idea of universality or common interest, facilitating precisely the opposite—pluralism—may be evidence enough of the dissimilarities.

Communicative Action and *Between Facts* addressed not only some of the major criticisms of *Structural Transformation* but also developed positive prescriptions for democratic society, revisiting the link between the public sphere and government. In *Between Facts*, Habermas makes an explicit claim for the structure of a democratic society. Whereas in *Structural Transformation*, the public sphere was (empirically) homogenous and was rather simply related to the sphere of government, in *Between Facts* Habermas's prescriptions are forced to address the fact of plurality in modern society. Furthermore, between writing *Structural Transformation* and *Between Facts*, Habermas developed his system/lifeworld dichotomy in *Communicative Action*, which led to the development of his idea, in *Between Facts*, of a sluice-gate mechanism to prevent power from the administrative and economic system infiltrating lifeworld contexts of interaction and social reproduction.

In moving beyond the Marxian class dichotomy, Habermas (1987) argues that the lifeworld, which is "a reservoir of taken-for-granteds, of unshaken convictions that participants in communication draw upon in co-operative processes of interpretation" (124), struggles against overextensions of the systemic imperatives of money and administrative power. Both lifeworld and system[4] attempt to coordinate society, but on Habermas's analysis, only the lifeworld can make a *legitimate* claim to social coordination. Habermas refers to the process whereby, through legislation and subversion of communicative rationality, the system penetrates forms of lifeworld sociation as the "colonization of the lifeworld." To better understand the difference between lifeworld and system, it is necessary to understand the types of rationality upon which they are based. Simply put, the lifeworld is based upon communicative rationality, coincidental with the "original" mode of language, whereas the system is based upon instrumental rationality, coincidental with a mode of language that is "parasitic" upon the original. The lifeworld relies upon, and is generated and sustained by, human communication, whereas the system does not. Habermas refers to the aforementioned process of colonization when instrumental rationality "surges beyond the bounds of the economy and state into other, communicatively structured areas of life and achieves dominance there at the expense of moral-practical and aesthetic-practical rationality." Systemic colonization doesn't go as far as to *replace* action oriented to mutual understanding; rather, it *disempowers* it: Steering mechanisms weaken communicative action's "validity basis so as to provide the legitimate possibility of redefining at will spheres of action oriented to mutual understanding into action situations stripped of lifeworld contexts and no longer directed to achieving consensus" (Habermas 1987: 304–11).

Whereas in certain contexts relieving the lifeworld of coordinating capacity is appropriate, in others, Habermas argues, it has a damaging effect. Whereas communicative action "offers the possibility of rationally motivated consensus," which is akin to how Habermas saw communication in the bourgeois public sphere, we in fact become empirically motivated in interactions motivated by money. The media of power and money "encode a purposive rational attitude. . . . and make it possible to exert generalized strategic influence on the decisions of other participants while bypassing processes of consensus-oriented communication . . . [T]he lifeworld is no longer needed for the coordination of action" (Habermas 1987, 183). Further to this, administrative power makes normative claims on citizens, which means that unless these requests are to be reduced to simple imperatives relying on sanctions, they must be legitimated (Baxter 1987, 59–61). On this view, the citizen's role in a colonized society has become "neutralized," there has been "a cleansing of political participation from any participatory content" (Habermas 1987, 350).

In *Between Facts*, Habermas constructs a system whereby such pathologies noted here can be avoided. In doing so he introduces his center (or core)/periphery dichotomy for the public sphere. The core is made up of "complexes of administration," the judicial system, and democratic opinion- and will-formation (in a parliamentary body). This nexus differs from the periphery insofar as the former has the capacity to act and is subject to formal rules.[5] On Habermas's normative account, the public sphere is a "*social space* generated in communicative action" that must be protected from systemic imperatives by separation. The periphery public sphere must be grounded in a civil society made up of "those *non-governmental and non-economic* connections and voluntary associations that anchor the communication structures of the public sphere in the society component of the lifeworld," and enables problems perceived in private life spheres to become amplified in the public sphere (italics mine). For Habermas, civil society has an egalitarian and open structure that mirrors the "essential features of the kind of communication around which they crystallize" (Habermas 1996, 366–67), that is, around communication oriented to mutual understanding, the inherent telos of human speech. On this account, the periphery public sphere itself is an elementary "social phenomenon . . . [that] cannot be conceived of as an institution and certainly not as an organization . . . [nor] a framework of norms with differentiated competences and roles, membership regulations and so on. Just as little does it represent a system. . . . the public sphere can best be described as a *network for communicating information and points of view*" (Habermas 1996, 360; italics mine). The public sphere is, like the lifeworld, "reproduced though communicative action." It is grounded in the lifeworld and leaves the "specialized treatment" of "politically relevant questions" to the political system. The political system should then only act upon issues that have been contested in the autonomous public sphere. Only after the latter process has taken place "can the contested interest positions be taken up by the responsible political authorities, put on the parliamentary agenda, discussed, and, if need be, worked into legislative proposals and binding decisions" (Habermas 1996, 314). The informal public sphere must be able to have an effect on the political system, but the latter must not adversely effect the autonomy of the former, lest decisions reached within the political system lack legitimacy. When this happens, Habermas argues that "the political system is pulled into the whirlpool of legitimation deficits and steering deficits that reinforce one another" (Habermas 1996, 386).

Indeed, in the sense of the *informal*, as opposed to the eighteenth-century bourgeois, public sphere, perhaps the Internet may act as a facilitating mechanism. If Habermas's requirement of the informal public sphere is that it "has the advantage of a medium of *unrestricted* communication" whereby it is more adept at perceiving problem situations, widening the discourse community, and allowing the articulation of collective identities and need interpretations (Habermas 1996, 308; italics mine), then there must be a medium to facilitate this. Importantly, the role of the media in the public sphere has been accounted for by Habermas. His claim in *Social Transformation* was that mass media was anti-democratic, duping the public into accepting manufactured opinion as their own. However, in *Between Facts*, Habermas had rejected much of the "cultural dope" approach to media studies, arguing instead that citizens adopt strategies of interpretation against media messages. Although he is by no means uncritical of the mass media in today's Western democracies, he seems to rely on media that simply lack the communicative capacity to facilitate the informal public sphere, leaving a communicative deficit. Even in its ideal form, removed from the imperatives of advertisers and with a strong public service ethos, broadcasting, for example, is still unable to facilitate effective communication in the informal public sphere. While it is better suited to the core public sphere, it will still suffer the problems of facilitating autonomous unfiltered communication, of translating informal messages, and of choosing the language (and direction) in which to translate such messages. In the following I intend to show

how the Internet is currently being utilized by groups that have the intention of filling this communicative gap.

The Informal Public Sphere, New Social Movements, and the Internet

A form of lifeworld activity in modern democracies is that undertaken by new social movements. New social movements (NSMs) patrol the boarders between the system and lifeworld, protecting the "grammar of ways of life," and also protecting civil society from encroachments by the system (Habermas 1987; 391–96; 1989b; 66–67; 1996; 373). In addition to this, NSMs generate collective identities, knowledge, and information. They are, as Eyerman and Jamison (1991; 55) assert, "like a cognitive territory, a new social space that is filled by a dynamic interaction between different groups and organizations." New social movements are precisely the bodies that perceive problems and push them onto the public agenda. NSMs on this account aim to generate and publish information that is generated autonomously from the needs of administration and the market. The loose structure of the Internet also provides for NSMs being, on Eyerman and Jamison's analysis, processes in formation, being the product of a series of social encounters. Traditional news and information services are simply not suited to such needs, being, as they are, reactive rather than active. Furthermore, NSMs may be regarded as taking the "moral point of view"; that is, they are not working for the interests of their members, but for those of humanity, or nature, as a whole.

NSMs have often faced hostile representations in the mass media. Numerous accounts of the latter (see Curran and Seaton 1991; Garnham 1992; Golding and Murdoch 2000; Hall 1982; Herman and Chomsky 1994; and Murdoch 1982) have illustrated the many structural barriers that prevent fair representation to "aberrant" groups. It may be said that traditional mass media are inadequate representative media not only because of social, political, and economic constraints, but also, in some respects, because of technological limitations that ensure that, even under ideal circumstances, conditions conducive to the reflexive interaction and information storage appropriate to social movements are not met.

To avoid accusations of technological determinism provoked by this comment, we can conceive of technological limitations in terms of the extent to which a medium or a technology can be shaped by its use depends upon its openness to influence. In this instance, the mode of development of the Internet was one in which multidirectional communication and cooperative activity played a central role. From as early as when former head of the Information Processing Techniques Office at the Advanced Research Projects Agency (ARPA), Joseph Licklider, penned "The Computer as a Communication Device," the importance of cooperation in the development of the networked computers became clear. Indeed, he specifies a particular view of how effective communication facilitates more effective research: "[S]ociety rightly distrusts the modelling done by a single mind. Society demands consensus, agreement, at least majority." He adds that "a particular form of digital computer organization . . . can improve the effectiveness of communication among people so much as perhaps to revolutionize" it. With the adoption of the Request for Comments (RFC) system,[6] which "inspired an open discussion model for creating common standards by consensus, with no barriers, secret or proprietary content" (Beckett 2000; 13–15), cooperation became semi-institutionalized and, arguably, reified into the application-structure of the Internet. The Internet Architecture Board (IAB, formerly Internet Activities Board), which is the main technical coordinating body, uses the RFC system. Before 1992,[7] the IAB regarded itself as "the coordinating committee for Internet design, engineering and management. The IAB is an independent committee of researchers with a technical interest in the health and evolution of the Internet" (RFC 1160, Section 2). Its functions were to set Internet standards, manage the RFC publication process, review the Internet Engineering Task Force (IETF) and Internet Research Task Force (IRTF), engage strategic planning, act as a technical policy liaison and representative to the Internet community, and resolve technical issues that cannot be treated by IEFT or IRTF. The standards procedures of the IAB are

> intended to provide a fair, open, and objective basis for developing, evaluating, and adopting Internet standards. They provide ample opportunity for participation and comment by all interested parties.

> At each stage of the standardization process, a specification is repeatedly discussed and its merits and failings debated in open meetings and/or public electronic mailing lists, and is made available for review via world-wide on-line directories. (RFC 2026, 1.2)

In adopting standards, the ultimate goal is to reach a considered consensus:

> [A]s much as possible the process is designed so that compromises can be made and genuine consensus achieved, however there are times when even the most reasonable and knowledgeable people are unable to agree. To achieve the goals of openness and fairness, such conflicts must be resolved by a process of open review and discussion. (RFC 2026, 6.5)

It can be seen, then, that organization of the technical bodies that direct development of the Internet mean that the latter can be considered to be open to a range of interests. Indeed, such concepts can be seen as constitutive of "Internet culture," or "computer culture." For example, the Free Software Foundation (FSF) was founded 1985 as a response to the continued commercialization of software and supposed decline of the original computer culture. Before the 1980s, the "computing community" did not regard the concept of "free" software, as that was all there was. Interestingly, as Richard Stallman points out, the term "hacker" originally referred to those who would alter programs and systems to improve them, the derogatory sense of the term only arising with the nondisclosure agreements forced upon users by new commercialized software. These nondisclosure agreements, essentially meaning that the user was not allowed to alter or share software, meant that the "cooperating community was forbidden," as hardware manufacturers forced for-profit software on users (Stallman 2002). Since it was founded, the FSF has a clear ethos that embraces the self-help and non-commercial "computer culture." The ethos of self-help, information generation and dissemination, and open public discussion is therefore far from alien to the computer and Internet "technocrats." It is this sense of culture of which many of the engineers, enthusiasts, and users are aware. Indeed, the IAB calls for the ethos of the Internet standards process, as outlined earlier, as having a significant role in shaping the culture of Internet users, urging them to embrace it "as a major tenet of Internet philosophy" (RFC 2026, 1.2).

That NSMs have taken to making good use of the Internet is not only theoretically consistent, insofar as NSMs seem to have a similar culture but empirically clear by the number of Web sites they have produced, as well as the range and extent of content. There are numerous forms of political, and therefore NSM, activity on the Internet. These include what Resnick (1998, 55–56) has termed "political uses of the Net," "politics within the Net," and "politics which impacts upon the Net." NSMs have engaged each of these forms. It can be seen that NSMs may *use* the Internet to support external activity, they may *work within* the Internet to create a foundation for their activities, and they have attempted to influence policy effecting the Internet. The anti-globalization, or pro-democracy as the Gramscians among us might recommend, movement has used the Internet very effectively, with the Association for Progressive Communications (APC) facilitating the former, while working within the Internet and attempting to influence policy.

If the lifeworld is premised upon human communication, and NSMs are perceived to be actors that support the former, then a communication system that is suited to NSMs must, to some degree, be supportive of communicative rationality in relation to the lifeworld. Following this logic, one must surely be justified in making the argument that in strengthening the lifeworld, the Internet can be seen as a foundational[8] medium for civil society and the informal public sphere. In particular, the Internet, with its global reach, could be said to be of value to social movements. The Internet enables social-movement groups and organizations to communicate, to generate information, and to distribute this information cheaply and effectively, allowing response and feedback. This is in large part because of its structure as a decentered, textual communications system, the content of which has traditionally been provided by users. Again, such characteristics accord with the requisite features of NSMs: nonhierarchical, open protocols; open communication; and self-generating information and identities. Further, the openness of Internet standards procedures to the admission of interests, as well as the open nature of discussion, means that Internet technology is indeed open to shaping by such groups in the pursuit of such ends. Although the Internet does have novel technological assets, for it to be a foundation medium, as it is well placed to be, requires appropriate interests to be sufficiently articulated and acted upon. The APC is concerned with doing so insofar as it is concerned with using the Internet to empower civic, social, and political movements; it is concerned with regulation; and it takes an interest in the actual structure and governance of the Internet. The work of the APC illustrates both the extent and the limits of radical action in using and shaping a technology. I will now turn to an analysis of the work of the APC to show what the extent and limit is.

The APC started out in 1990 as the first "globally interconnected community of ICT users and service providers working for social and environmental justice" (APC Web site), formed by various NGO and civil society networks. The APC currently has twenty-five member networks serving more than fifty thousand activists, nonprofit organizations, charities and nongovernmental organizations (NGOs) in over 133 countries. The APC is committed to supporting international links with member and partner networks from Western, Central and Eastern Europe, Africa, Asia and the Pacific, Latin America, and North America. They are one of the largest Web-based NGO/civil society organizations in the world, and they are truly global. It might be argued by the most judgmental of observers that the APC has a Western "bias." This is true insofar as one would argue that democracy, freedom, and civil society are Western notions (a complex argument that I don't want to get into here), because APC promotes a (again arguably) Western technology, and because it is registered in California. On the whole, however, the APC is a prime example of how foundational organizations can facilitate NSMs on the Internet. The APC's mission statement is worth citing at some length to illustrate the extent to which its self-perception fits with a Habermasian conception of the role of NSMs:

> The Association for Progressive Communications is a global network of non-governmental organizations whose mission is to empower and support organizations, social movements and individuals in and

> through the use of information and communication technologies to build strategic communities and initiatives for the purpose of making meaningful contributions to equitable human development, social justice, participatory political processes and environmental sustainability. (About APC 2002)

It can be seen that the APC is interested in laying the foundations upon which other groups can build. The APC aims to empower others to make "meaningful contributions to . . . participatory political processes," or, in fitting the Habermasian model, the APC aims to strengthen the communicative capacities of the lifeworld, enabling it to assert itself against systemic imperatives. As well as developing its own Internet charter (http://www.apc.org/english/rights/charter.shtml), the APC's *Internet Rights Working Group* worked with GreenNet (the UK member) to develop the GreenNet Civil Society and Internet Charter further spelling out its objectives, including facilitating

> the right to communicate [which] should be recognised as a fundamental right for everyone. In a modern society in which communications has assumed a central role, those that cannot be heard become largely ignored. It is essential for democracy that such exclusion be ended. New communications technology must be made available to all. (GreenNet 2002)

In addition to this fundamental claim, APC asserts its opposition to censorship, and promotes the protection of privacy and the pursuit of open democratic processes in setting Internet standards and developing technologies. With regard to the latter, the APC has seen high levels of success. A recent example of how the APC has been involved in the pursuit of democratic procedures in Internet standards is with regard to the new governing bodies. As noted, technology and the interests that shape them change over time. Indeed, this can be seen concretely in the changes underway in the governing structures of the Internet.

In the mid 1990s, a series of reforms took place that resulted in the privatization of the physical structure of the Internet. Accordingly, the constitutions of the governing bodies were altered. Whereas it is doubtful that the Internet in any of its manifestations ever saw a "golden age" of self-organization away from the influence of industry and government, it is surely the case that the balance of power has been altered. Such a point has been related to me by a number of engineers involved in the original APRA program. However, whereas such engineers emphasize the plurality of interests that influence the direction of the development of the Internet, it is clear that the dominant systemic interests have changed from those of the government to those of the commercial sector. Thus, the mission statement of the now defunct Internet Assigned Numbers Authority (IANA) was that it was "dedicated to preserving the central coordinating functions of the global Internet *for the public good*" (IANA 2002; italics mine). However, the public good ethos of IANA seems to have disappeared with the "privatization" of the Internet. As a result, the Internet Corporation for Assigned Names and Numbers (ICANN), whose mission is to "facilitate the coordination and management of only those specific technical managerial and policy development tasks that require central coordination" (ICANN 2002), replaced IANA.

However, there is a struggle within ICANN to reform its mission. The struggle for the definition of the roles and responsibilities of ICANN itself was manifested in a recent Bertelsmann-funded conference on the role of ICANN in Internet governance. As noted in Marcel Marchill's (2001) *Recommendations for Internet Governance* on behalf of the Bertelsmann Foundation, there are at least two differing interpretations of ICANN (44). On one hand, from the perspective of the U.S. government, ICANN regulation is in private hands, financing is internationalized, the status of ICANN as a nonprofit organization is trivial, and the need to legitimize it is very limited because the mandate is *purely technical*. On the other hand, "from the point of view of numerous stakeholders outside the United States and critics from both within and without," ICANN is an American organization whose authority is dependent on the United States, whose legal status is guaranteed by Californian law, and which lacks legitimacy. Further to this, proponents of the latter view are not passive in their belief that ICANN's mission should be political and social rather than purely technical. At the conference for which these recommendations were drafted, there were clear conflicts between the directors of ICANN, who were divided roughly along these lines. It is clear that the U.S. government view is problematic from the outset. Not only do domain name options presuppose particular legal regimes and entities, and indeed prioritize some over others, but this process has become explicitly formalized in the Domain Name Disputes Resolution Policy (DNDRP). The DNDRP is a classical example of how, to paraphrase Webster (1995), the Internet is shaped by "real-world" economic and social relations. The APC sees such fundamental struggles as central to its mission. It sees its role as shaping the Internet, rather than taking its technological or cultural basis for granted. For example, the APC itself has been involved in the struggle for the soul of ICANN. In the recent election for the board of directors of ICANN,[9] the APC recommended and supported candidates, and succeeded in getting three elected. On the back of this success, and in accord with their Civil Society and Internet Charter, the APC aim is to form ICANN in accord with the needs of civil society. This approach is formalized in the APC-backed "Civil Society Statement on ICANN Elections," which expresses a similar conception of civil society to that of Habermas. In this, they clearly distinguish civil society from the market, in contradistinction to classical liberal conceptions, as well as the state in accord with Habermas's own normative account:

> Civil society is a third sector of society alongside the state and the market. The values underlying civil society include freedom of association, freedom of expression, participatory democracy, and respect for diversity. A vigorous civil society is an important counter-balance to government and business. (Civil Society Internet Forum, 2002)

Moreover, the APC's account of the development of Internet culture is similar to the aforementioned, in that "when neither commerce nor governments paid too much attention to the Internet, the people setting the standards worked within a prevailing Internet culture favouring openness and the consensus of all stakeholders." These presuppositions cause the APC's view of ICANN to contrast with that of the U.S. government noted earlier: "Technical coordination of the Internet's core resources has unavoidable social, economic, and political consequences" (APC Issues in Internet Rights 2002).

The successful election of APC candidates to the board of ICANN has important implications for the future of the Internet. This success is even more notable in consideration of the former's record in assisting social movements to date. In terms of its practical support of new social movements, the APC has been rather successful. Perhaps their most notorious project was their support of the southern Mexican rebel organization, the Zapatistas. The Zapatistas have been referred to by Manuel Castells (1997) as the first informational guerrilla movement. However, it is not the case that the Zapatistas are engaged solely in an information war. Indeed, their struggle against repression from the Mexican state is very real, very material, and very violent. The Zapatistas took to arms in the early 1990s partly as a response to systemic threats (such as the NAFTA agreement) to their lifeworld. In the course of their struggle for social justice, they recognized the common struggle of all humanity,[10] incorporating a very wide range of interests, and thus interest, into their movement. Notwithstanding the material base of their struggle, in certain respects Castells is correct in his assertion. The Zapatistas became rapidly adept at utilizing counterinformation to publicize their cause, assisted by NGOs and social movements with appropriate computer equipment.

In turn, the well-known case of the Chase Manhattan Bank memo illustrates the successful implementation of Internet technologies by the Zapatistas and their supporters. Written during the peso crisis of December 1994, the report called for the elimination of the Zapatistas as a means to convince international financiers and speculators that the Mexican government was in control. The document was leaked, originally appearing in a limited circulation paper newsletter that was by and large ignored. When the story was published on the Internet, however, it circulated so widely and so quickly that it was picked up by the mainstream media, and the resultant negative publicity and widespread protests forced Chase to disassociate itself from the report and from its author. In fact, such was the impact of the APC's support of the Zapatistas that it prompted the Rand Corporation to pay special attention in their analysis of the Zapatista movement:

> Of these (outside network organizations), the most important from a technological and training standpoint is the Association for Progressive Communications (APC), a global network of computer networks that has many affiliates. . . . The APC and its affiliates amount to a worldwide computer-conferencing and e-mail system for activist NGOs. It enables them to consult and co-ordinate, disseminate news and other information, and put pressure on governments, including by mounting fax-writing and e-mail campaigns. The APC also helps activist NGOs to acquire the equipment and the training their members may need in order to get on-line. (Ronfeldt and Martinez 1997, 10)

This serves as a useful example of upward pressure from civil society to government and the economy.

Since the various successes of the APC in their influence upon the governing bodies, and in facilitating the fulfilment of social movements' information and communication requirements, they have gone further in attempting to shape the application structure of the Internet. To this end they have developed ActionApps, employing the General Public License of the FSF (see APC, "ActionApps. Building Information Communities"), to further help NSMs. ActionApps are, essentially, software that enables persons to update Web-based information without specific skills in Web site design. Furthermore, ActionApps also enable users and groups to pool information and share resources, thus extending their body of knowledge and information, which, in particular, accommodates Eyerman and Jamison's conception of social movements as being intimately tied to knowledge production, broadening the range of knowledge past that which suits business and administration. Indeed, ActionApps have enormous benefits for social movements, but only time will tell their long term effectiveness and their impact upon the application structure of the Internet.

Reservations

As the earlier account shows, the APC has been relatively successful not only in developing the Internet infrastructure itself, but also in assisting the practical activities of social movements. So, if one rejects the view that the Internet can be thought of as a public sphere in itself, we might come to regard it as a supporting foundation upon which public spheres can be built. There are, however, still some reservations about this capacity, not only in terms of the APC's use of the Internet, but also in the structure of the latter itself. In the first case, the APC has minimal interactivity in their own Web site, but this may not be as important as it first appears because of to the fact that their primary objective is to enable other groups. Furthermore, the degree to which they do facilitate other groups is considerable. A notable contribution of the APC to the communication structures of social movements is their development of the aforementioned ActionApps.

Perhaps a more pressing problem, as it effects *all* Internet users, is that of the communicative capacity of the Internet itself. As I noted earlier, lifeworld or civil-society phenomena must have a communicative capacity. Yet this capacity must not be thought of as simply providing voice. For *communicative* action to be successful, there are a number of requisite conditions. Without going into too much detail on this complex matter, Habermas takes communicative action to be premised on the existence of criticizable validity claims. That is, whenever we act communicatively, we raise claims that the other party(s) in communication can question. In order for a speech act to be accepted, the hearer must be able to accept its truth, the corresponding normative basis, and the sincerity of the speaker. Of course, such criteria might be unattainable on the Internet. First, verification of information on the Internet, as with any medium, is a complex process requiring the will and time that many are not prepared to invest. Second, the lack of a shared lifeworld, or even a shared cultural background of international Internet users,[11] causes problems for the acceptance of normative acceptability or rightness. Finally, the much-championed anonymity of the Internet makes the assessment of sincerity very difficult indeed. As the difficulty of ascertaining truth is common to most media, I want to focus briefly on the latter two points.

In the first case, that of normative rightness, it is clear that there is not a set of norms freely developed by an international civil society from which participants in Internet communication can draw. In his empirical study of Usenet newsgroups, Wilhelm (1999) found that the content of communication was "dissonant, unmoored to contemporary language norms" (162). However, one only need think of Netiquette to realize how rapidly norms rise from chaos. Further, Slevin makes the point that the Internet is able to "connect up many cultures and different experiences, increasing the likelihood of clashes of interest" because "given the interactive potential of Internet technology, such views are *easily* challenged and revealed for what they are" (2000, 196; italics mine). It might be suggested that such a situation is analogous to what Habermas referred to as the risky freeing up of language from convention. Mere convention becomes weakened through Internet activity as the exposure to alternatives and the need to explain and justify directly to the Other means "the need for reaching understanding is met less and less by a reservoir of traditionally certified interpretations immune from criticism. . . . [The lifeworld] can be regarded as rationalized to the extent that it permits interactions that are not guided by normatively *ascribed* agreement but—directly or indirectly—by communicatively *achieved* understanding" (Habermas 1984, 340).

Anonymity on the Internet presents itself as a double-edged sword in terms of communicative action. On one hand, it means that those perhaps too shy or otherwise inhibited will feel more confident about expressing their opinions, especially opposing political opinions against someone who might be physically intimidating. Further to this, profession, class, accent, body language, gender, ethnicity, religiosity, physical stature, speech impediments, and so on all act as potential obstacles to "real-world" face-to-face discussion, but are not as apparent online. Thus Professor X must rely solely on the strength of his or her argument against shop assistant Y, so we might say that in such cases, "consensus formation rests *in the end* on the authority of the better argument" (Habermas 1987, 145). However, although Slevin (2000) asserts that "the paradox of isolation and visibility means that those using the Internet . . . must still treat distant others on peculiar terms of equality" (185), in practice the *responsibility* to do so is not immediately apparent. Thus, on the other hand, one might insist that anonymity means that the fundamental requisite of human communication, responsibility, is lacking in Internet communication.[12] In this sense, a discussant may simply disengage in debate, with his or her anonymity making it impossible to be compelled to continue. In addition to this, a discussant may disengage, simply to reengage under a different identity. Therefore, with anonymity comes irresponsibility, and responsibility is one of the most important, yet perhaps most underemphasized, aspects of Habermas's theories of communicative action, discourse ethics, and public sphere. A precondition of communicative action taking place is that actors accept responsibility for their utterances. Indeed, this is the basis upon which speech act theory rests and from where many of Habermas's critics begin. The balance of anonymity and responsibility often depends on what sort of discourse is sought. On one hand, if the background culture of the user is authoritarian (to whatever degree), then anonymity is an important tool that enables criticism without the fear of repression. On the other hand, if there is a liberal political culture, the likes of which exist only in approximation, then anonymity loses its role as security, leaving the question of whether anonymity serves to allow utterances to carry only their internal weight at the expense of responsibility.

Conclusions

In the course of this chapter I have shown first that a more comprehensive understanding of the work of Habermas enables one to gain a better understanding of how the Internet works in relation to society, and the democratic importance of social movements in this process. Furthermore, I suggest that the drawbacks and benefits of the Internet and the Web are not to be simply understood—there are complex relations between society, the Internet, government, business, Internet governance bodies, and political and social movements. What the Internet can do for these agents is not predefined in the technology itself, but is open to definition by the users, and both citizens and systemic steering media are struggling for hegemony. In this sense, the APC can be seen as a significant movement in attempting to secure not only the use but also the structure of the Internet against systemic imperatives.

In view of this conclusion, it is imperative that Internet users take an interest in how they shape the medium, especially its communicative capacity. Perhaps the most pressing challenge for Internet users, one that Habermas's work can also illuminate, is that of commercialization and control, of colonization. As I have alluded to at various points in this chapter, there has been a recognizable shift in the content, use, and structure of the Internet over the past five years. This has occurred as business and government have began to take more of an interest in what they can gain from this medium. The former have seen money-making opportunities and have pushed governments to secure the Internet for their own use, and the latter have seen the propaganda, surveillance, and administrative potential of the Internet. The development of the Web has been an important factor in this process by increasing the number of "consumers," by enabling multimedia presentation of goods, and by making the navigation and use of the Internet so much easier.

Whereas on one hand the Web has allowed greater access to a greater number of communication technologies than before, on the other it might be said to have reduced the interactivity of the Internet as official political and business Web sites are developed to act as one-way propaganda platforms.[13] Whereas it would be absurd to use Usenet as a one-way communication mechanism, it is now becoming acceptable to use the Web in such a way. So, rather than the users themselves providing the majority of content on the Internet, companies and governments are colonizing more and more. In fact, one might argue that a form of enclosure is occurring whereby "small-holders" are being forced into the heavily populated, controlled, and regulated areas such as those provided by America OnLine and Microsoft Network. If this process continues Internet users will be increasingly herded along predefined enclosures, or channels, which become more and more difficult to leave, rendering the Internet just another colonized mass medium providing standardized information and discussion, limited interactivity, and everything the consumer needs to satisfy her or his manipulated material desires. Again, this process, as grand as it may seem, is not predetermined. It is up to citizens, representatives, and political, social, and cultural movements to stake their own claims on the frontier and ensure they remain protected as necessary.

Notes

1. See, for example, Mark Poster's (1995) well-known essay, Friedland (1996), the collection of essays in Toulouse and Luke (1998), Malina (1999), Wilhelm (1999), and Slevin (2000).

2. See Dahlgren (1995) for an account of his employment of four methodological dimensions: media institutions (or media structure), media representations, social structures within which media operate, and social interaction in terms of reception.

3. Although e-mail is probably the most commonly used Internet application, the prevalence of Web-based e-mail means that the www is increasing as a primary access point to e-mail outside the workplace.

4. For simplicity's sake, I use "system" to refer to the economic and administrative subsystems.

5. This capacity to act differs in terms the organizational complexity: "the parliamentary complex is the most open for perceiving and thematising social problems, but it pays for this sensitivity with a lesser capacity to deal with problems in comparison to the administrative complex" (Habermas 1996, 355).

6. RFCs are documents that articulate problems faced by engineers and invite responses. The problems may be technical in essence, but RFCs also include more "political" questions. They are open to all for contributions and are published in full, for all to see, on the RFC Web site, www.rfc-editor.org.

7. In 1992, the Internet Society (ISoc) was formed as a result of the broadening of the Internet community and the "need for community support" (Leiner et al. 2000) in processes of governance. The ISoc is now regarded to be the main governing institution for the constitutive structure of the Internet.

8. I use the term "foundational" here in the sense of the foundations, the core, of civil society. As John Keane (1998) has asserted, there are certain prerequisites for civil society that must be strengthened and maintained.

9. Elections took place in 2001. However, at the time of writing this chapter, there is a discussion relating to the future of a democratic ICANN. Some directors have expressed concern as to the effectiveness of the elections. For example, the Africa ICANN member was elected with only 67 of a total of 130 votes cast in Africa. Others, such as board member Andy Mueller-Maguhn, point to a corporate takeover of ICANN. This is a useful example of the limits of influence. Indeed, if the procedures and the substantive resources are skewed against civil society, then the outcome is to an extent predetermined.

10. "All humanity" is my term. More specifically, Subcomandante Marcos, the leader of the Zapatistas, argued for "indigenous brothers and sisters, workers, peasants, teachers, students, farmworkers, housewives, drivers, fishermen, taxi-drivers, office workers, street vendors, gangs, the unemployed, journalists,

professionals, nuns and monks, homosexuals, lesbians, transsexuals, artists, intellectuals, sailors, soldiers, athletes and legislators, men, women, children, young people and old, brothers and sisters" (*The Guardian* 15th March 2001) to join his struggle.

11. This is contingent on the extent to which one accepts that we are sufficiently "globalized" at the moment.

12. The relationship between anonymity and responsibility is a complex one. Indeed, it might be the case that those who habitually frequent particular Web sites—especially the kind of new social movement sites traditionally dedicated to ideals of transparency, honesty, and so forth—are more likely to conform to such principles. I am grateful to Shivdeep Grewal for proposing this point. Furthermore, IP blocking software can be used to ensure that "flaming" and other such abuses do not prevent proper debate taking place. It must also be noted, however, that such software is often used to block legitimate comments that do not accord with those of the Web site, newsgroup, or bulletin board. Further to this, the question of responsibility in "real-world" discourse has engaged moral philosophers for millennia.

13. The World Wide Web Consortium (W3C) is the main body for the agreement of www protocols and standards. This body is fast becoming the most important in terms of questions of e-commerce, privacy, property "rights," and communicative capacity.

References

Adorno, T., and Horkheimer, M. 1997. *Dialectic of Enlightenment*. London: Verso.

APC. "ActionApps. Building Information Communities." Online. http://www.apc.org/actionapps/english/general.

——. 2002. "ICANN Election Results." Online. http://www.apc.org/english/rights/governance/icann_election2000/index.shtml.

——. "APC Internet Rights Charter." Online. http://www.apc.org/english/rights/charter.htm.

——. 2002. "Issues in Internet Rights—Governance of the Internet." Online. http://www.apc.org/english/rights/governance/.

——. 2002. "About APC." Online. http://www.apc.org/english/about/indexs.html.

Barbrook, R., and Cameron, A. 2002. "The Californian Ideology." Online. http://cci.wmin.ac.uk/HRC/ci/calif5.html.

Baxter, H. 1987. "System and Lifeworld in Habermas's Theory of Communicative Action." *Theory and Society* 16: 39–86.

Beckett, D. 2000. "Internet Technology," pp. 13–46 in *Internet Ethics*, ed. D. Langford. London: Macmillan Press.

Calabrese, A., and Borchert, M. 1996. "Prospects for Electronic Democracy in the United States: Rethinking Communication and Social Policy." *Media, Culture and Society* 18: 249–68.

Calhoun, C. 1992. *Habermas and the Public Sphere*. Cambridge, MA: MIT Press.

Castells, M. 1997. *The Power of Identity*. Cambridge, MA: Blackwell.

Civil Society Internet Forum. 2002. "Civil Society Statement on ICANN Elections." Online. http://www.civilsocietyinternetforum.org/statement.html.

Curran, J. and Seaton, J., eds. 1991. *Power without Responsibility: The Press and Broadcasting in Britain*. 4th ed. London: Routledge.

Dahlgren, P. 1995. *Television and the Public Sphere: Citizenship, Democracy and the Media*. London: Sage.

Dutton, W. 1996. "Network Rules of Order: Regulating Speech in Public Electronic Fora." *Media, Culture and Society* 18: 269–90.

Eyerman, R., and Jamison, A. 1991. *Social Movements: A Cognitive Approach*. Cambridge: Polity.

Friedland, L. 1996. "Electronic Democracy and the New Citizenship." *Media, Culture and Society* 18: 185–212.

Garnham, N. 1992. "The Media and the Public Sphere," pp. 359–76, in *Habermas and the Public Sphere*, ed. C Calhoun. Cambridge, MA: MIT Press.

Golding, P., and Murdoch, G. 2000. "Culture, Communications and Political Economy," pp. 70–92 in *Mass Median and Society*, ed. J. Curran and M. Gurevitch. London: Arnold.

GreenNet. 2002. Civil Society and Internet Charter. Online. http://www.gn.apc.org/action/csir/charter.html.

Habermas, J. 1971. *Towards a Rational Society*. London: Heinmann.

——. *The Theory of Communicative Action: Reason and the Rationalization of the Lifeworld*. Cambridge: Polity Press.

——. 1987. *The Theory of Communicative Action: The Critique of Functionalist Reason*. Cambridge: Polity Press.

——. 1989a. *The Structural Transformation of the Public Sphere*. London: Polity Press.

——. 1989b. *The New Conservatism: Cultural Criticism and the Historians' Debate*. Cambridge, MA: MIT Press.

——. 1996. *Between Facts and Norms*. London: Polity Press.

Hague, B., and Loader, B. 1999. *Digital Democracy*. London: Routledge.

Hall, S. 1982. "The Rediscovery of 'Ideology': Return of the Repressed in Media Studies," pp. 56–90 in *Culture, Society and the Media*, ed. M. Gurrevitch, T. Bennett, J. Curran, and J. Woollacott. London: Routledge.

Herman, E., and Chomsky, N. 1994. *Manufacturing Consent: The Political Economy of the Mass Media*. London: Vintage.

Hoff, J. 2000. "Technology and Social Change: The Path between Technological Determinism, Social Constructivism and New Institutionalism," pp. 13–32 in *Democratic Governance and New Technology*, ed. J. Hoff, I. Horrocks, and P. Tops. London: Routledge.

IANA. 2002. Online. www.iana.org.

ICANN Fact Sheet. 2002. Online. http://www.icann.org/fact-sheet.htm.

Keane, J. 1998. *Civil Society: Old Images, New Visions*. London: Polity Press.

Leiner, B. et al. 2000. *A Brief History of the Internet*. Online. www.isoc.org/internet/history/brief.shtml.

Licklider, J. 2002. "The Computer as a Communication Device." Online. http://memex.org/licklider.pdf.

Malina, A. 1999. "Perspectives on Citizen's Democratisation and Alienation in the Virtual Public Sphere," pp. 23–38 in *Digital Democracy*, ed. B. Hague and B. Loader. London: Routledge.

Marchill, M./Bertelsmann Foundation. 2001. *Who Controls the Internet? Recommendations.* Berlin: Bertelsmann Foundation.

Mawhood, J., and Tysver, D. 2000. "Law and the Internet," pp. 96–126 in *Internet Ethics*, ed. D. Langford. London: Macmillan Press.

Murdoch, G. 1982. "Large Corporations and the Control of the Communications Industries," pp. 7–26 in *Culture, Society and the Media*, ed. M. Gurrevitch, T. Bennett, J. Curran, and J. Woollacott. London: Routledge.

Poster, M. 1997. "Cyberdemocracy: Internet and the Public Sphere," pp. 201–17 in *Internet Culture*, ed. D. Porter. Routledge: London.

Resnick, D 1998. "The Normalisation of Cyberspace," pp. 48–68 in *The Politics of Cyberspace*, ed. C. Toulouse and T. Luke. London: Routledge.

RFC 1160. 2002. "The Internet Activities Board." Online. ftp://ftp.isi.edu/in-notes/rfc1160.txt.

RFC 2026. 2002. "The Internet Standards Process–Revision 3." Online. ftp://ftp.isi.edu/in-notes/rfc2026.txt.

Ronfeldt, D, and Martinez, A. 1997. "A Comment on the Zapatista 'Netwar'," pp. 369–91 in *Athena's Camp: Preparing for Conflict in the Information Age*, ed. J. Arquilla and D. Ronfeldt. Washington, DC: Rand Corporation.

Slevin, J. 2000. *The Internet and Society*. London: Polity.

Stallman, R. 2002. "The GNU Project." Online. http://www.gnu.org/gnu/thegnuproject.html.

Toulouse, C., and Luke, T. 1998. *The Politics of Cyberspace*. London: Routledge.

Weckert, J. 2000. "What Is New or Unique about Internet Activities?" pp. 47–64 in *Internet Ethics*, ed. D. Langford. London: Macmillan Press.

Webster, F. 1995. *Theories of the Information Society*. London: Routledge.

Wilhelm, A. 1999. "Virtual Sounding Boards: How Deliberative Is Online Political Discussion?" pp. 154–178 in *Digital Democracy*, ed. B. Hague and B. Loader. London: Routledge.

Winston, B. 1998. *Media Technology and Society: A History from the Telegraph to the Internet*. London: Routledge.

"Zapatistas March into the Heart of Mexico." 2001, March 15. *The Guardian*.

6

Comparing Collective Identity in Online and Offline Feminist Activists

Michael D. Ayers

Someone might say that I have made love to a good portion of the participants. Is there anything more exciting than yelling NOWWWWW!!!! during climax?

—NOW Village member

Is this the activist identity one has in an online feminist group? I interviewed two sets of feminist activists: one set participates in the online NOW Village,[1] which was started by the National Organization for Women, and the other set participates in an offline activist organization called Womanspace. I wanted to see if the members of the online activist group could have the same sort of collective identity that the group meeting face-to-face would have.

The social-psychological concept of collective identity has been used by social-movement theorists to explain how a social movement can maintain and build strength over time. In this chapter, I discuss my comparative research of two feminist activist groups in the context of the theory of collective identity, specifically asking what kinds of collective identity are possible in cyberspace.

Studying Feminist Activists Online and Offline

Feminists' presence in cyberspace spans the long-standing feminist discussion sites Cybergrrl, geekgrrl, Nerdgrrl, and Homegrrrl to the creation of more than two distinct gender categories in the LambdaMOO multi-user domains sites[2] (see Bell 2001 for a discussion of gender and cyberspace). I chose to focus on a "mainstream" feminist online group, the NOW Village, because other online feminist activist sites were less active and/or were not established activist organizations.[3] NOW has been an active social-movement group since the 1960s. They have incorporated Internet technologies into their activist framework and have provided space through their Web site to foster a group discussion about feminist issues. The NOW Village group members used a Web-based textual discussion board where registered participants can post messages about whatever they want. This group is made up of primarily women who are involved in the group on either a daily basis, semi-daily basis, or weekly basis. Some men are participating in this group, but no men were used in this study.

At first glance, the NOW Village looked like a full-fledged social-movement group. When you enter into the main NOW homepage, it has defining features that represent the organization as a social-movement organization (see Figs. 6.1 and 6.2). In the middle of the page are news alerts that are updated on a fairly regular basis. These news alerts describe world or national affairs that have an impact on NOW, either in a positive or a negative way. Across the top of the page, an image of women and men of different races joining together to march for NOW hold signs with slogans displaying what the collective stands for (see Fig. 6.3). The point that comes across is this: These people are together, as a group, fighting for a cause. Below this "unifying" image, we can see links to various parts of the NOW site: everything from how to support NOW and how to contact NOW chapters that might be in one's regional area to how to get technical support. So this appears to be the place where someone interested in issues that NOW supports should come.

Along the left side of the page, the user is invited to click on various links for various ways to become involved *immediately*. One can sign up for e-mail action alerts and receive updates about certain topics with which NOW concerns itself, such as lesbian rights and reproductive

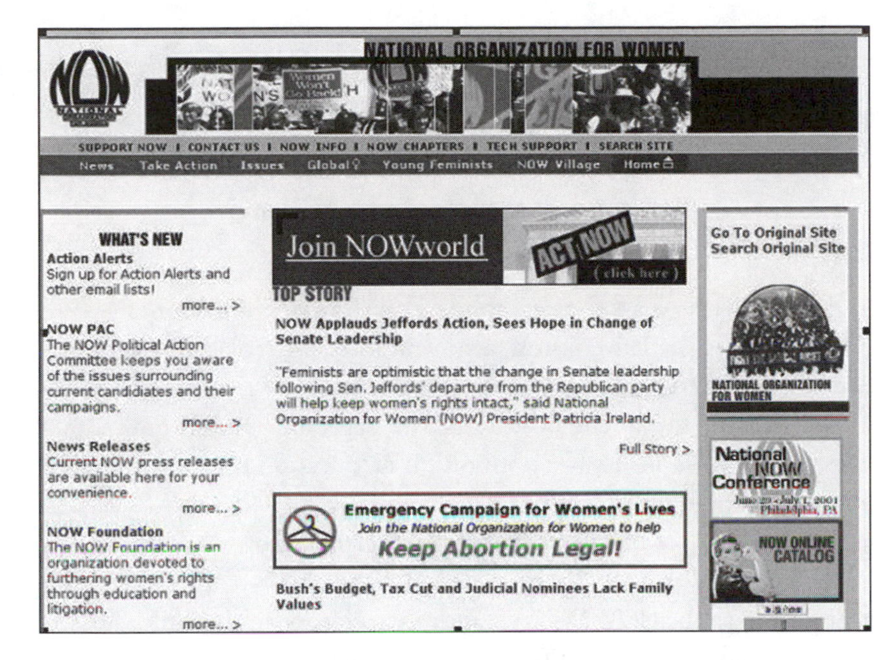

Figure 6.1 · NOW homepage top half (www.now.org).

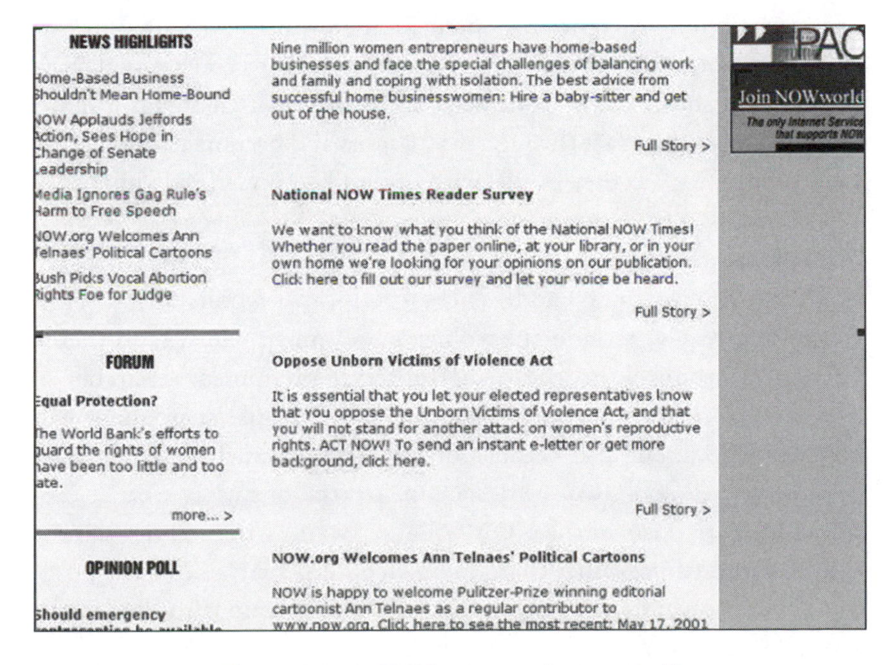

Figure 6.2 · NOW homepage bottom half.

Figure 6.3 · NOW homepage banner.

rights. The user has choices about how he or she becomes active. Further down the left column, a potential (or current) activist can be led to current NOW press releases or back issues of the in-print NOW newsletter. Finally, at the bottom of the page, an opinion poll allows the user to voice his or her opinion on a "question of the week" that reflects the organization's concerns.

All of this seems essential for a movement group to attract new members through its Web page. The link bar at the top of the page contains a link to the NOW Village. Once a user clicks on this link, the person is transported into a different part of NOW's cyberspace as well as something that is completely different from NOW. This time, it is a little different from the usual user/text interaction. The unifying banner of people coming together for change remains at the top of the Web page, indicating this "village" is a space for people to come together in the name of social change. But what is different is that this is a space where a user can interact through discussions and become part of a "virtual community" (as they put it) with other like-minded individuals.

Once a user or potential user enters the NOW Village discussions, one can immediately sense a group feeling. People address each other by their "screen names," address the group as a whole, and discuss things that revolve around the women's movement, NOW, and themselves as a group. So to the casual observer or trained researcher it appears that there is a group here, not only because it is under the NOW umbrella, but also because of this daily interaction. But as the old saying goes, looks can be deceiving. To find out what kind of collective identity exists in the NOW Village, I wanted to interview NOW Village participants, rather than doing a content analysis of their postings. Interviews allow me to compare the online group with the offline group, the members of which I also interviewed.

In the spring of 2001, I conducted nine in-depth qualitative semi-structured interviews with two groups of feminist activists—those who work mainly online and those who work mainly face-to-face (f2f). I considered the online feminist activist group as such because they use the Internet to foster both activist-related relationships and social change through discussion over the Internet. I considered the offline feminist activist group as such because they are dedicated to feminist social change and their presence in cyberculture is limited to e-mailing one another over a listserv to announce f2f organizational meetings and, occassionally, events and issues that will be discussed at future f2f meetings.

To obtain my sample of NOW Village interviewees, I contacted group participants who had provided an e-mail address and who were active participants in the group. I contacted only those with fifty or more total posts and who had been active in the group for one month or more. To participate in the NOW Village, one must be registered through NOW (registration is free). However, anyone who can access the Web site can observe the members' discussions. The group discussion board also gives a registered user or nonuser access to an information box on the specific person. This information box allows a person to see the date that the participant registered, the total number of posts they have contributed, an e-mail address (if they gave one), a homepage address (if they gave one), and their occupation, geographic location, and/or interests.

I interviewed five members of the NOW Village group, contacting them by e-mail and inviting them to participate in the research project. I contacted ten people and of the ten, five responded and participated in this study. I interviewed all five over e-mail (because of geographic distance and participant preference) in a back-and-forth e-mail exchange. A total of thirty-three e-mails were exchanged with the NOW Village respondents.

I interviewed a comparison group of four activists in a local feminist group known as Womanspace. This is one of two feminist activist groups that are located in the Montgomery County, Virginia, area. This group operates at a local university and is concerned with consciousness-raising and activism around women's reproductive rights, the problem of sexual assault, and other feminist issues. This group holds f2f meetings regularly where they plan community and regional activism. To get the respondents from Womanspace, I attended one of the meetings and asked for people willing to be interviewed for this project. Seven people indicated that they were interested, so I contacted them by e-mail and/or phone and arranged face-to-face interviews. Of these seven, four responded and granted interviews for this study. I conducted the interviews in a mutually agreed-upon space on the campus. Each interview lasted approximately forty-five minutes.

I had typed responses from the interviews with the online group and tape recordings and field notes from the interviews with the offline group. My interview questions focused on uncovering aspects of a collective identity. I inquired about the activists' relationships with other members of the group, activity levels, and the types of social-movement activities in which the group participates. All of my questions were designed to get at the key features of collective identity: shared definitions, levels of consciousness, boundary markers, and negotiations that social movement groups make in the private sphere on how they will present themselves in the public sphere.

My online interviews with participants in the NOW Village[4] reveal three themes: sexuality, personal gain, and opposition. In contrast, my interviews with the offline group, Womanspace, reveal themes of collective identity, what social-movement scholars suggest makes for effective political activism. Before discussing these themes and their implications for a collective identity among my interviewees, I use the social-movement literature on collective identity to explain the four components of collective identity.

Collective Identity: Toward a Working Definition

The concept of collective identity helps social-movement scholars explain why a person would want to participate in a movement when he or she is seemingly satisfied with his or her current economic state. Melucci (1996) defines collective identity as "an interactive and shared definition produced by several individuals (or groups at a more complex level) and concerned with the orientations of action and field of opportunities and constraints in which the action takes place" (44). For

Melucci and others, collective identity requires both an interactive group and shared definitions (J. Gamson 1995; W. Gamson 1992; Melucci 1989, 1994, 1996; Taylor and Whittier 1992). Shared definitions of reality and of right and wrong help a person link her beliefs to the larger group's same belief, thus attaching the individual to the group. These cognitive definitions must be concerned with the group's action and the larger society in which the group is situated (Melucci 1996). Therefore, cognitive definitions reflect a movement's group feelings and directly reflect the action in which the actors participate.

Like Melucci, Taylor and Whittier (1992) agree that collective identity involves a group's shared definition about its situation and place in the larger society, but they go on to identify three additional characteristics of a collective identity: boundaries, consciousness, and negotiation (105, 109). For a social movement group, boundaries mark off the group from an opposition by emphasizing differences between the actors in the group and the opposition (111). Usually, this "marking off" is done by a dominant group in society; boundaries are set to distinguish the minorities in a society from what is shared and held to be "normal" under the dominant group belief (111). Therefore, boundary markers are central in collective identity formation because they stress the minority groups' shared perceptions as being distinct from the dominant group. Boundary markers range from an ascribed status, such as race, to differences about what is culturally right in our society, such as gay marriage rights (see J. Gamson 1997).

In addition to boundary markers that locate a group's place relative to other groups, a level of consciousness of the actors within the group is required for a collective identity to become established (W. Gamson 1992; Melucci 1989, 1994, 1995; Taylor and Whittier 1992). This means that a group becomes aware of itself through a series of self-reevaluations of shared experiences, shared opportunities, and shared interests (Taylor and Whittier 1992, 114). Taylor and Whittier also note that this consciousness is apparent in formal documents, speeches, and writings that the group shares.

Finally, subordinate groups use a process of negotiation to build their collective identity (Taylor and Whittier 1992, 118). Privately, a group will negotiate new ways of thinking and acting in the public sphere. Taylor and Whittier (118) describe "identity negotiations" in which the group involves itself in direct or indirect ways of freeing itself from the dominant institution or culture. In their study of lesbian feminist collective identity, Taylor and Whittier describe how the lesbian feminist communities renegotiate what it means to be a "woman," both privately and publicly (e.g., rejecting traditional notions of what it means to act, dress, and look "womanly").

We can combine these two definitions (Melucci 1989, 1994, 1995; Taylor and Whittier 1992) to make a working definition of collective identity for use in this study. Collective identity is an interactive and shared definition system that incorporates boundary markers, consciousness, and complex levels of negotiation to situate the individuals and the group in the larger arena of a dominant-subordinate belief system. I turn now to my interview results with this understanding of collective identity in mind.

Collective Identity in Cyberspace

After conducting interviews with NOW Village members and Womanspace, I found that orientations to the political group were markedly different across the online and offline groups. In this section I discuss these differences and their implications for collective identity in the online group. Because my interest is in whether or not online political groups can achieve the collective identity required for effective social-change work, I discuss the online group and use the offline group for comparison.

Sexuality

Four of the five NOW Village respondents explained that political issues related to sexuality were reasons for being involved in the movement and in the NOW Village. However, while sexual freedom and sexual tolerance were discussed as being very important to the individual and were reasons for being involved in the women's movement in general, these were not cited as reasons for participating in the NOW Village specifically.

Participants could be using this group to connect with others for sex. A good example of this can be seen in this response from Carrie:[5]

> Someone might say that I have made love to a good portion of the participants. Is there anything more exciting than yelling NOWWWWW!!!! during climax? My involvement in NOW has been for less than five years, but I have been involved in the battle for the rights of women for a long time. One of the things that women, even lesbians, can do is seducing members of the religious right and helping to break up their families. Is there anything more exciting than seeing two crying children of a pastor after his wife has divorced him after he has had sex with a woman such as myself? Do I do that kind of thing? I could be accused of doing that any time that I get the opportunity. That sort of thing can be soooo exciting. And some pastors are quite good in bed, though I would much rather have sex with a woman.

Carrie also posted the following to the NOW Village this spring, reiterating these ideas in a topic that she started around the issue of "skin mags":

> Why should there be a problem with them. Is not nudity and the sex act very natural acts that should be able to be enjoyed by all? As long as this sort of thing is made to sound dirty, will our society ever be truly free? At one time, women voting was considered to be bad. At one time, freeing the slaves was considered to be bad. At one time, being a Lesbian was considered to be bad. At one time, premarital sex was considered to be bad. Does morality seem to be an absolute to anyone? As long as it is fun, enjoyable, and does not hurt anyone, what is wrong with buying, selling or looking at Skin mags?

Here it seems that she is equating her participation in the movement and with NOW around issues of sexuality and power. When I asked Carrie to describe any bonds or ties that she felt with the other members of the NOW discussion boards she replied with: "Unless you mean things such as the joy of fisting publicly, why should I have bonds with anyone that I am not intimate with. Without intimacy, what good is a relationship." Another example of this can be seen with this statement from Maggie when I asked her if she considered herself involved in the current women's movement:

> Of course I am. Why ask a question like that? Why would anyone question my involvement. People like me need to be involved with fighting for women. I fight and fight and fight. I am emotionally involved, mentally involved, and volitionally involved. This is real life and I have NEVER EVER slept with a woman who is not on my wavelength.

When asked about the strategic planning or real-world activism that might be planned in the NOW Village, Sharon replied, "I have not found much [activism planned]. This is not to say that three or four of us have not exchanged nude pictures of each other." When asked about any group ties or bonds that might be present within the NOW Village, Sharon also replied with, "If you want to know about my sex life, why should I engage in that kind of talk here? Sexual dynamics I guess would have some similarity. I like to be openly sexual both with a woman and on the board."

So for these people who participate in the NOW Village, sexuality seems to be an issue, but there is no indication that they are participating in this group because the other group members have similar viewpoints on sexual freedom or how to go about changing sexuality-based inequalities.

The interest in sexuality shown by the participants without any accompanying interest in the feminist social-change work of the broader NOW organization suggests that their involvement with the group revolves around online flirtation and/or meeting potential sexual partners. To be sure, sexual activities occur in plenty of social-movement groups. For example, former leader of the Black Panthers Elaine Brown (1992) portrays a vivid account of the sexual encounters while in that group. But whereas that sexual activity was a by-product of the Black Panthers' struggles for civil rights, it appears to be a main purpose among the NOW Villagers that I interviewed. It is also equally important to note that with any qualitative research answers to given questions must be taken at face value. I do acknowledge that some of these answers do sound cliché and could be meant to "shock" the researcher, especially since I am a man researching feminist activists.

Personal Gain

A characteristic of being a member of a social-movement group dictates that participants are working for social change: the collective-action results are more important than any personal gain that might result from participation. By personal gain I mean having mainly selfish reasons for participating. Reports of personal gain through the use of these boards came up in all five interviews. As I have already explained, collective identity emerges through group goals and accomplishments, not through individual goals and accomplishments.

These respondents are having an experience different from people in typical social-movement groups. Four out of five respondents reported using the discussion group for specifically "picking up" other women. When asked to describe the relationship that one has with people who are online who participate in the NOW Village, Laura replied with, "I have not picked up any lesbian lovers but I am developing relationships"–indicating that having personal sexual gain through this participation is a reason for participating in the NOW Village. When asked the same question, Maggie replied with: "I have one sexual online relationship with a woman who is attractive and with whom I have enjoyed spiritually and sexually. Overall I have had my good and bad relationships online." Again, we can see that when asked about relationships with people that are seemingly participating in an online social-movement group, Maggie did not think of these relationships as being built around the movement, but instead thought of personal, sexual relationships.

That one user identifies herself as "blueyed" provides further evidence that the respondents use the group for personal gain. (I expected to find online handles like "sojourner," "grrlpower," or "pro-choice.") When asked to describe how the use of these discussion boards makes them feel in terms of their personal activism, blueyed responded with: "I just use it for practicing in arguing." It was not determined who blueyed argues with, whether it is in-group feminists or oppositional users (discussed later). A level of conflict can be expected in any group setting, as can sexual relationships, but these are usually a by-product of group participation, not the raison d'etre.

In contrast, all four interviews with the offline group, Womanspace, yielded discussions about Womanspace as a positive place for building friendships and growth. This social-movement group provides a place for like-minded individuals to come together and support each other in the causes that they believe in, and as a result, close-knit friendships form because of activism. We see this clearly in a statement by one of the respondents: "I would definitely say that Womanspace is really close. You know, the majority of my closest friends here at this university are involved in Womanspace. And although we are a really close group, I don't feel that we are exclusive, either." One other example that depicts the feelings of closeness within this group came from a respondent who said, "I feel accepted because there are lots of mutual understandings between all of us." This feeling of closeness creates a strong bond; as such it is a quintessential sign of a collective identity.

All four of the offline respondents discussed the importance of friendship in attracting new members. The friendly environment that seems to be fostered within this group positively effects the collective identity of the group. This points to friendship as a way that collective identity can be sustained in a group.

Opposition

All of the NOW Village respondents in this study cite an opposition. Some are vague references to opposition in general and some are direct references to opposition that has arisen within the NOW Village. The literature says that for a strong collective identity to exist, contact with a defined opponent should bring the group closer together. All of the respondents acknowledge an oppositional presence—trolls—*within* NOW Village. When asked to discuss the opposition that one encounters when using the NOW Village, blueyed replied:

> There is plenty of that [opposition]! I think the opposition is actually a coordinated group of individuals. I've seen examples of them subtly correcting others or "reminding" them of things that seem to be from off-line. I think they are very careful with the words they choose and then pick out things other than the issues to attack with. For example, it is easy for them to lash out at "feminists" while maintaining that they are not misogynists. Women don't have the luxury. There is

no single word for "men that have improper views of women." ("Misogynist" doesn't work because they can claim that they are married and love their wife so they can't hate women.)

From January through April 2001, I witnessed an increase in people joining the group for the sole purpose of causing trouble by trying to discuss issues that are counter to the movement or the group's ideas. For example, one of the most active trolls is a person who is constantly quoting the Bible and offensive pro-life views. This will obviously stir people into arguments, since one of the principles that NOW organizes around is pro-choice. But, NOW Villagers don't really see this opposition as a threat to the group or the movement in general (which in offline contexts would be a threat).

When asked to discuss the opposition that occurs within this group, one respondent said that they are "mostly fools for what I have seen. Some people have written a lot of words, but have not said a whole lot." Another respondent summed up the difference between opposition that she faces in real life versus when involved in the online group with the simple statement: "Real life opposition is much worse. But real life sex I find to be much better in my life."

NOW Villagers distinguish real life from online life. Because there is no physical presence of opposition, it seems that there isn't really much of a threat of actual opposition within the group. At the same time, the inability to keep outsiders out of the online conversations may make the activist ties weaker. A question that this research raises is, How should we actually define a social movement group in an online setting? If it allows for communication on opposite sides of the movement, like NOW Village does, is it still a social movement group? Or is it some sort of new, different social movement group that has really yet to be defined? If offline activity is considered "real" and online activity is considered not real, then the online opposition would not help to create a strong collective identity because it would not be perceived as truly threatening. I will explore this more a little later when I discuss group negotiations.

Boundary markers are culturally constructed barriers that mark a social group as being distinct from the rest of society. In a social-movement context, this usually occurs because the group has viewpoints that are not congruent with the larger society. Through boundary markers the group's collective identity solidifies because the group will highlight the ideas and thoughts that members have in common. I asked questions of the NOW Village participants about the opposition that these group members faced, either online or in real life. The presence of the online opposition that I described earlier is drawing a boundary online and thus creating a distinction between who is for the group causes and who is not.

The literature on boundary markers cites a defined opposition as key to creating a distinction between the social-movement group and others. All of the online interviewees discussed an opposition that is present, but as I said before, they don't really see it as anything to worry about. There is a definite distinction that these respondents make between online opposition and opposition in the larger movement context. Concerning this, Maggie said, "Real life opposition is like the life that I live. Online opposition is more like doing 'research' and turning in silly little papers to overeducated fools who have sheets of parchment on their walls." To clarify this, I asked whether she felt real-world opposition is basically incomparable to online opposition and whether she felt that online opposition is something that is somewhat of a joke, to which she responded, "I think that you may be understanding me."

All of the respondents from Womanspace discussed opposition that their group faces as a general attitude that certain parts of society have about sexual assault, rape, and women's rights in general. Examples of these sectors could be any group that is promoting patriarchical values. One of Womanspace's goals is to change a general attitude that these sectors seem to have: that feminists (such as Womanspace members) are aggressive and negative people. They feel that the best way to go about creating change is through peer education and consciousness-raising. One respondent described this when she said, " I think that there is a negative connotation about what feminism is. We are viewed as aggressive for some reason."

Womanspace is a principal organizer of Take Back the Night, a march that seeks to raise awareness on sexual assault issues in the community and on campus. When asked about what she sees as the general public's perception of Womanspace's efforts, Lorrie said, "People don't realize what it's about and I don't think people want to know." Margo echoed this idea when she said, "The best step to change is to promote visibility in the community and advancement through peer education." This theme of opposition that has emerged seems to be having a positive effect on the group's collective identity by uniting them with a cause and creating something to work for. Like other conventional social-movement groups, this one is no different in terms of having a defined opposition that is separate from the group and what the group is trying to change.

Shared Definitions

Shared definitions–what a group defines as right and wrong in our society–can be determined if a number of respondents' answer with the pronoun "we," indicating a collective or shared definition or belief. The NOW Village group respondents in this sample do not seem to have shared ideas about what is right and wrong. When asked about the goals that the group is working toward, all of the respondents referred to their individual goals of participation. There were no "we" statements that indicate a collective identity or that the online group is acting as a collective. Instead of saying "we try to get people to think" blueyed said, "I try to get people to think." Out of all the statements that would indicate evidence of shared definitions, not one respondent discussed anything that would be considered ideas shared by the group. Instead their reasons for participating in this group were more individual and personal. These individualistic responses question the actual existence of the NOW Village as a social-change group. Even though there were no reports of any specific shared definitions that the group has, the fact that there is cited opposition toward the group is a shared definition that can be identified. These participants are defining what is right and wrong by way of identifying the opposition and their opposing viewpoints.

The respondents from Womanspace seem to have two shared definitions that are helping to create a collective identity. First, three of the four respondents discussed the notion of equality not only for women, but also for gays and lesbians. Another important shared definition that this group has is the prevalence of sexual assault in society and the idea that this is a major problem in our society and specifically the local community. We can see evidence of this with Lorrie's statement, "Speaking as a representative of Womanspace, I would say one of our major focuses on this campus are sexual assault issues because that's a real problem on our campus."

Levels of Consciousness

Levels of consciousness are the group's self-awareness through shared experiences, shared opportunities, and shared interests. My interviews reveal no real group consciousness in the NOW Village but a group consciousness in Womanspace. When asked about any feelings that are perceived when interacting and participating in the NOW Village (with members who share similar view points), blueyed summed it up best: "I think there is a lot of frustration. Part of the down side to the boards is that people don't get the human face to face stories." Later, when asked if she felt any "bonds" or "ties" with other members of this group, she responded, "I don't feel any bonds or ties with other members. I think real groups try and work out more practical solutions. These boards don't encourage that." This response indicates the necessity of f2f interaction to become a group, work as a group, and relate to one another as a group.

Turning to Womanspace, when asked if she felt close to other group members, Erica replied, "Yes, I do feel a closeness because you can't go through the events we do and not [feel that way]." Margaret replied, "It sort of becomes a very important part of your life; you work together on things and get to know each other and it sort of turns into a social circle. Everyone has a closeness to one another."

Womanspace organizes and attends protest marches, including locally organized marches, regional marches, and national marches. For example, every January the group goes to Lobby Day in Washington, DC. These experiences create an awareness that in turn solidifies through a reciprocal relationship a strong collective identity based around equality, women's rights, and sexual assault prevention.

Negotiations

Social-movement groups discuss among themselves how they plan to create social change. When asked about any strategy sessions or "plans of attack" that the group might discuss for creating change (e.g., planning marches, protests, or letter-writing campaigns) the general reaction on the NOW Village can be summed up with Carrie's response: "Are you shitting me? Not much if any that I know of." One respondent did mention some planning by some members to attend an upcoming march, but this does not seem to be an everyday occurrence. At least according to the other four respondents, no strategic planning occurs. This raises the question: Why would one person report planned activism and others not? Taking into consideration the small sample size, this still raises the issue of the solidarity that this group has, and also whether this is a social-movement group at all. As discussed earlier in this chapter, negotiations are ways that a group will discuss in the private sphere how it is going to present itself in the public sphere. Usually this is how the group plans strategies for creating change.

Womanspace works constantly in the private sphere to promote change and get its message across. One example of this is brainstorming sessions that occur twice a month and are used for general planning. Margaret said, "If we are planning an event, we'll all have input: who do we want to come, what do we want to happen, and one person will take the lead on using everybody's ideas. It's pretty well balanced: the people who volunteer to take things on."

The negotiation processes that take place are helping to create the collective identity of this group. Group members interacting with one another is an important way that the group creates a collective identity. According to the responses from this sample, negotiations are an important part of Womanspace.

Directions for Future Research on Online Collective Identity

The research presented in this chapter calls into question the nature of what comprises a social-movement group in cyberspace. Forming a collective identity in cyberspace may be difficult—although hopefully not impossible—because of the distance between group members. Although my study offers just one particular group comparison, further research is necessary to find out whether or not other activist groups that have been created solely online are more successful than the NOW Village.

It is important to remember that these respondents are a very small fraction of the overall women's movement, and their reflections and opinions cannot represent the beliefs of anywhere near a majority. Furthermore, the respondents from the NOW Village are just five people from one organization that has over 550,000 members in the United States. The answers that they have given are not reflective of this organization's overall goals; they are only five sets of beliefs at one given point in time. Thus, no generalizations about NOW or the broader women's movement, online or offline, can or should be made from these interview data.

Does the existence of a group of people operating online under an activist group umbrella necessarily mean they are an activist group? The online group members I interviewed did not seem politically and socially motivated outside of the confines of their computer screen. An online social-movement group must have some level of activism in the "real" world if the changes it seeks politically are to go beyond the realm of the Internet itself. Research into online political groups must clarify what counts as activism.

A final issue raised by this research has to do with the applicability of traditional social-movement theory to online activism. Is collective identity required for online social-change action? At least in the case of the NOW Village members I interviewed, the identity of feminist activist meant online sex, not online social-change work. My offline interviewees from Womanspace would surely not consider that real activism. This chapter raises questions about how traditional notions of collective identity formulate in cyberspace. Scholars investigating online activism will have to decide whether or not now-current theories of social movements apply to online political groups, and possibly update the literature to reflect how, if at all, social movements work online.

Acknowledgments

I would like to acknowledge and thank to no end: Martha McCaughey for indescribable inspiration and countless reads and re-reads, Rachel Parker-Gwin for inspiration and motivation when I was very impressionable, and the Lower East Side of New York City.

Notes

1. When I revisited Now.org a year later in the spring of 2002, I was saddened to learn that NOW Village no longer exists and quickly realized, as well as questioned, the power that organizations have to create and uncreate groups in cyberspace, such as the NOW Village.

2. For starting points on research and theorizing cyberfeminism, see *Women@Internet* (Harcourt 1999) and *Wired Women: Gender and New Realities in Cyberspace* (Cherny and Weise 1996).

3. I chose to investigate the social-movement groups that began on the Internet, first investigating Spiderwomen.org and Women Leaders Online. However, these had too little activity on their discussion lists and Web sites. The public archives of the Spiderwomen.org group listserv revealed that little to no discussion was taking place between the months of August 2000 and January 2001. This caused me some alarm because it seemed odd that a social-movement group discussion list would be dormant during a heated presidential election year. I thus deemed the group inactive and ineligible for study. Similarly, I decided that the Women Leaders Online group was ineligible for study because their Web site was not updated. Furthermore, to obtain access to the group's listserv, one has to pay to become a member of the group. I did not go this route, but instead used the group homepage's infrequent and uninformative updates to discern that this group is also not an active social-movement group. Thus I finally chose the online extension of NOW, the NOW Village, to study because of its regular online activity.

4. By heavily involved, I mean they make frequent posts and are engaged in discussions almost daily.

5. All of the names in this chapter have been changed to protect the identities of those I interviewed.

References

Bell, David. 2001. *An Introduction to Cybercultures*. New York: Routledge.

Brown, Elaine. 1992. *A Taste of Power: A Black Woman's Story*. New York: Anchor Books.

Cherny, Lynn, and Elizabeth Reba Weise, ed. 1996. *Wired Women: Gender and New Realities in Cyberspace*. Seattle, WA: Seal Press.

Gamson, Joshua. 1995. "Must Identity Movements Self-Destruct? A Queer Dilemma." *Social Problems* 42: 390–407.

——. 1997, April. "Messages of Exclusion: Gender, Movements, and Symbolic Boundaries." *Gender and Society*: 178–99.

Gamson, William. 1992. "The Social Psychology of Collective Action," pp.

53–76 in *Frontiers in Social Movement Theory*, ed. Aldon D. Morris and Carol McClurg Mueller. New Haven, CT: Yale University Press.

Harcourt, Wendy, ed. 1999. *Women@Internet: Creating New Cultures in Cyberspace*. London: Zed Books Limited.

McAdam, Doug. 1988. *Freedom Summer*. New York: Oxford University Press.

Melucci, Alberto. 1989. *Nomads of the Present: Social Movements and Individual Needs in Contemporary Society*. Philadelphia: Temple University Press.

——. 1994. "A Strange Kind of Newness: What's 'New' in New Social Movements?" pp. 3–35 in *New Social Movements*, ed. Enrique Larana, Hank Johnston, and Joseph R. Gusfield. Philadelphia: Temple University Press.

——. 1996. *Challenging Codes: Collective Action in the Information Age*. Cambridge, MA: Cambridge University Press.

NOW homepage. 2001. Online. http://www.now.org.

Taylor, Verta, and Nancy Whittier. 1992. "Collective Identity in Social Movement Communities: Lesbian Feminist Mobilization," pp. 53–65 in *Frontiers in Social Movement Theory*, ed. Aldon D. Morris and Carol McClurg Mueller. New Haven, CT: Yale University Press.

7

Mapping Networks of Support for the Zapatista Movement

Applying Social-Networks Analysis to Study Contemporary Social Movements

Maria Garrido and Alexander Halavais

The structure of global communication has been undergoing a quiet sea-change. It was once reasonable to assume that communication among those in different nations would naturally be channeled through a hierarchy of institutions, through patterns that had been established over centuries and at great social cost. Although there were certainly examples of decentralized global grassroots organizations, these were placed at a significant disadvantage because of the logistical apparatus available to states and to corporations.

In the sweep of a decade the situation has changed drastically. The most widely cited example of the way that the new dynamics of social networks interplay is the Zapatista movement in the southern state of Chiapas in Mexico. On January 1, 1994, an army of about three thousand indigenous peasants united under the banner of Ejército Zapatista de Liberación Nacional (EZLN) took up arms and occupied seven towns in Chiapas (Schulz 1998). This uprising was provoked by an urgent need to fight together against the extreme poverty that had deterred the social and economic development of indigenous communities in Mexico.

The date the Zapatistas decided to take these towns by force was the same day that North American Free Trade Agreement (NAFTA) took effect. What makes the Zapatista movement unique from a historical perspective, and what makes it a model of participatory efforts toward social change, is its extensive use of the Internet as a tool for global mobilization. By January 3, 1994, two days after the uprising, Subcomandante Marcos—the figurehead of the movement—was online. Marcos became the first hero of the Net, and his "Lacandona jungle address became the locus of a global news agency whose dispatches were written by guerrilla combatants themselves" (Halleck 1994, 30).

The Zapatista movement has been called both a model social movement on the one hand and the first instance of Net warfare on the other, views that are both equally accurate (Ronfeldt et al. 1998). While it is widely recognized that the Zapatista movement has been particularly successful because of the networked nature of its effort, few have clearly charted what such a network might look like. The research presented here is an exploratory attempt to analyze the main characteristics of the global social networks of solidarity that support the Zapatista movement in cyberspace. The purpose of this study is to create a tentative map of the Zapatista network on the Web that can help us to illustrate some of the ties, roles, and strategic alliances that have been built within and around the movement worldwide. By doing this, we hope to understand the composition of the Zapatistas social network, as that network is reflected in hypertext structures. Given the central role of the Internet in the social structure of the movement, we argue that a map of network connections is, in effect, a map of the social and organizational relationships that constitute the most significant part of the Zapatista movement. A careful examination of this hyperlinked network of Web sites provides a unique insight into the character of the Zapatistas' phenomenal success, and particularly the degree to which the group has become a catalyst for a transnational network of activists.

New Communication Structures

Since the end of the cold war, power has been redistributed among actors that until recently had no significant presence in the international public arena. This redistribution of power, together with the development of new communication technologies, has led to a "reweaving of the political fabric of international and national dialogue, upsetting the traditional balance of power in the creation of domestic and foreign policy" (Cleaver 1998, 2).

Two factors have triggered this reorganization. The first and most important of these is globalization, which has fostered a relative decline of the power of states while nourishing the rise and strength of nonstate actors (Mathews 1997). Along these lines, De Angelis (2000) argues that the globalization of world trade and production has increased interdependency among international actors, and concurrently helped to draw together the needs and aspirations of a variety of hitherto separated groups and individual actors across the globe. Second, and in concert with the first factor, a telecommunications revolution has facilitated the exchange of information among underrepresented groups in society and has opened alternative spaces wherein these groups can make their voice heard by the international community.

Manuel Castells (1997, 68) argues that "the trends towards globalization and informationalization created by networks of wealth, technology, and power are enhancing productivity, driving cultural creativity, and increasing the communication potential" within a global civil society. Most importantly, they are setting the stage for a new form of collective action for the information age (Melucci 1996).

New decentralized communications networks have led to fissures in the international structures of power, fissures that have been exploited by new actors on the global stage (Mathews 1997). There has been a tremendous growth in cross-border networks among nongovernmental organizations (NGOs), including the hundreds that mobilized against NAFTA during the 1990s and those that gathered in Seattle to protest the secrecy of the World Trade Organization (WTO) in 1999 (Cleaver 1998; Mathews 1997). Such cross-national networks not only bypass national government policymakers but often work directly against their policies, particularly the so-called neoliberal reforms.

Fundamental to these efforts is the need of social movements to seek alliances with others and to make the struggles of other movements their own. The fight of one becomes the fight of all; their terrain of struggle transcends national boundaries and acquires instant global scope (Cleaver 1998; De Angelis 2000; Schulz 1998). Although the struggle against structural reforms and globalization has evolved since the 1980s, the development of information technology in the last decade has facilitated communication among members of this international network of social resistance. The effective use of communication networks has broadened the scale of action for these movements, empowering their struggle internationally and opening new spaces—what might be called "virtual publics" (Jones and Rafaeli 2000)—that move beyond the exchange of information to facilitate shared culture, coordination, and solidarity (Cleaver 1998; Schulz 1998). Many have noted that nascent virtual organizations supported by the Internet and related networks have the potential to become the vanguard of a technological globalization that will bring about a global citizenry (*Der Spiegel* in Froehling 1997; Rheingold [1993] 2000). These new networks represent a spirit of interaction that is unique; they "speak a language that seems to be entirely their own, but they say something that transcends their particularity and speaks to us all" (Melucci 1996, 1).

Why Map the Zapatistas' Online Network?

Some scholars argue that the EZLN has played no direct role in the proliferation of the use of the Internet. Rather, they argue, the efforts toward the building of the network of support in cyberspace were initiated, and actually are maintained, by others, particularly those in the western world that support the Zapatista movement (see Cleaver 1998). However, all agree that the Internet played a crucial role as a catalyst to disseminate information about the indigenous struggle in Chiapas around the world and opened the space for the creation of networks of transnational support, whether through direct use by the Zapatistas or through intermediary networks that existed primarily in computer-rich countries (Cleaver 1998).

Castells (1997) writes that the Zapatistas were unique for their use of information technology to build an international network of solidarity. As Schulz (1998) points out, global interactive communication has enabled the Zapatistas to link up with individuals, groups, and organizations—particularly in the industrialized world. This cross-national solidarity has been of crucial relevance to the Zapatistas' continued survival because it has encouraged international support of the movement, while at the same time strengthening their position with the Mexican government. Similarly, Cleaver (1998) argues that the evolving computer networks supporting the Zapatista movement are providing the backbone or nerve system for increasingly global opposition to the dominant economic policies of the present period. For this reason, he explains, it is not an exaggeration to speak of a "Zapatista Effect" (Cleaver 1998, 622).

If the structure of the Zapatista movement is what makes it unique, it is important to investigate it from a structural perspective, describing the movement in terms of the relationships among its constituent elements. Given the degree to which the Zapatistas have made use of the Internet, it represents a natural target for investigation. Hyperlinks provide a direct measure of relationships among documents on the World Wide Web, and possibly an analog for other structural relationships among the core Zapatista movement and other movements around the world.

Representatives within the Zapatista movement have made clear that their strategy is to exploit new communications technologies to create global relationships. The Zapatista movement encompasses a participatory process for social change, one that is concerned as much with social equality, freedom, and participation in decision-making as it is with economic opportunity, women's rights, and reduction of poverty in indigenous communities. These aims extend not only to Mexico or Latin America, but around the world. In the "First Declaration of Reality," Subcomandante Marcos states:

> The new distribution of the world excludes "minorities." The indigenous, youth, women, homosexuals, lesbians, people of color, immigrants, workers; the majority who make up the world basements are presented, for power, as disposable. The new distribution of the world excludes the majorities (quoted in De Angelis 2000, 23).

By diversifying the discourse of struggle, the Zapatistas have become an icon of social resistance and an example to follow for social change. Even though the movement has underscored its grass roots in the fight for indigenous rights, self-determination, autonomy, and cultural preservation in Mexico, its fight has become a call for justice and economic opportunity for all those underrepresented and exploited groups around the globe.

Another crucial strategy for the Zapatistas' effort toward increasing their network of support was organizing the International Encounter for Humanity and Against Neoliberalism, which took place in a small town called La Realidad in Chiapas, Mexico (Schulz 1998). This "intercontinental meeting" attracted thousands of activists who gathered in La Selva Lacandona hoping to open a multicultural dialogue and to form an international alliance to fight against the inequities of globalization and neoliberalism (Cleaver 1998; Froehling 1997; Schulz 1998). Schulz (1998) defines this strategy as part of the "communicative praxis" of the Zapatista movement, a process of constructing meaning, projects, visions, values, styles, strategies, and identities through deliberate engagement in dialogue both with supporters and detractors.

Cross-national solidarity facilitated by the use of the Internet has empowered and strengthened the Zapatista movement and has allowed its survival. For this reason, we argue that in order to understand the structure of this transnational social network, we must analyze the deep architecture of its online network of support. These networks are an unusual phenomenon; they are at the same time decentralized within the broader spectrum of the structure but they perform specific roles within their surrounding networks. Despite their specialization, the complexity of these social networks allows for an interaction far less influenced by differences in gender, class, or race than interactions in other media might be (Froehling 1997).

The Internet is being used at the grassroots level to promote international discussion and connections that link struggles and often bypass the nation-state. Cleaver (1998) describes three examples in which these interlinking movements are facilitating dialogue and are creating an alternative niche in cyberspace. First, he argues that the Internet facilitated the spread of information around the world about *indigenous experiences* in seeking alternatives to create a culturally, linguistically, and ethnically heterogeneous democratic sphere. These experiences, he notes, were successful at building networks among a diverse array of indigenous people at the local, regional, and international level. Second, Cleaver explains that the *environmental network* is another highly elaborated sphere in cyberspace. It links environmental movements with indigenous environmental practices. This relationship has allowed for a more developed explanation of the relationship between indigenous culture and the natural environment and a slow merging of these networks of communication. Finally, Cleaver discusses the inclusion of a *women's network*. Triggered by the drafting of the Revolutionary Women's Law by indigenous women, many women's networks have established direct connections with indigenous women in Chiapas and have played an active role in circulating information about the Zapatista movement.

Following a similar path, Markus Schulz defines the Zapatistas' "social network capacity" as one of the key elements for the success of the movement and spreading international support. In probably the most comprehensive study of the Zapatista movement from the network perspective, Schulz (1998) argues that their social network capacity has made the Zapatistas less reliable or dependent on their internal military organization than on the support they receive from individuals and associations that are explicitly not part of the EZLN. Furthermore, the Zapatistas proposed the formation of an Intercontinental Network for Humanity and Against Neoliberalism that was intended to create links of resistance and communicative access among and within the social actors of this international network. Schulz argues that globalized interactive communication has enabled the Zapatistas to link up with heterogeneous individuals and organizations, particularly in Western countries, that organize on behalf of the Zapatista cause. These have become crucial for the movement because they have bolstered their position with the Mexican government. The author concludes that the Zapatista insurgency can be thought of as a new type of transnational social movement emerging in the global order to counter globally defined threats and the shrinking of national political action spaces.

Indeed, this is precisely what the Zapatista movement has come to represent in the minds of many. Showing that its value is more than iconic, that there is evidence of a global structure, is far more difficult.

By examining closely the structures of communication that allow for the movement's message to spread—in this case via the World Wide Web—we can provide some indication of the extent and character of the organization's global involvement.

Social-Network Analysis and Hyperlink Analysis

Social-network analysis seeks to describe networks of relations, trace the flow of information through them, and discover what effects these relations have on people and organizations (Garton, Haythornthwaite, and Wellman 1997). Social-network analysis has emerged over the last century as a method of discovering patterns of exchange and relationships among groups. Early work can be found among social psychologists examining the emergence of "cliques" and among anthropologists concerned with systematically describing the structure of tribes (Scott 2000). Those interested in social networks have developed a set of tools, many adapted from graph theory, to help uncover and characterize these networks (Galaskiewicz and Wasserman 1993; Wellman and Berkowitz 1997).

With the advent of information technologies and computer-mediated communication, social-network analysis has seen a resurgence. As people make greater use of computer networks to fulfill social needs, these computing networks are themselves clear indicators of communication structures within a society. As Barry Wellman (2001, 2031) puts it, "computer networks are inherently social networks." We proceed, then, with the assumption that a map of the communication network is roughly isomorphic to the structure of relationships among the users (Garton, Haythornthwaite, and Wellman 1997; Scott 2000; Wellman 2001).

The network perspective seems ideal when studying newly networked organizations, and as we have seen many scholars take this overall perspective to help explain the structure of NGOs and their use of the Internet. Fewer operationalize this construct and examine the computer networks that undergird these larger social networks in anything more than a superficial way. A number of potential empirical applications of social-network analysis would help elucidate the structure of social movements, including those that map the connections between organizations (Diani 1992).

The World Wide Web provides a ready source of such networked information. Exchanges over e-mail and on listservs provide more dynamic information, but the World Wide Web has several advantages. Unlike other applications of the Internet, it is largely public and easily accessible. In addition, the structure of the Web, although changing, evolves far more slowly than other linkages might. As a result, we can obtain a "snapshot" of sorts of the current relationships between organizations based on the relationships between their Web sites.

When creating Web sites, site authors naturally tie their own efforts to allied ones through hyperlinks. Since establishing a hyperlink is a conscious social act executed by the author of a Web site, we may assume that some form of cognitive, social, or structural relationship exists between the sites. As Adamic and Adar (2001) entitled a recent paper, "You Are What You Link." Surveys of Web masters and other work indicate that hyperlinks represent reasonable approximations of social relationships (Jackson 1997; Kling 2000). While a complete picture of social networks cannot be drawn without knowledge of the content of communications, important descriptive work can be done using the structures of interconnection alone.

At a large scale, this Web of linkages provides an indication of the "landscape" of related movements. These structures, which are not apparent to the casual Web surfer, only come to light under an analysis of larger Web linkages. A number of attempts to describe this structure have appeared in the literature of various fields under the terms "hyperlink analysis" or "webometrics" (Adamic and Adar 2001; Björneborn and Ingwersen 2001; Brunn and Dodge 2001; Halavais 2000; Kim 2000; Kleinberg 1999; Park 2002). While social-network analysis gives us an established set of methodological tools from which to draw, how these measures relate to the Web remains an open question.

The questions at hand, then, are: How do Web sites related to the Zapatistas interact with a larger network of NGOs? Have they served as a catalyst for larger networks of NGOs? Several measures used in social-network analysis can be of help in addressing these questions. Given the size of the sample, the first step is categorizing the results into cohesive subgroups, based upon co-linkage structures. The approach is similar to that used in citation analysis of scholarly literatures. By examining these subgroups, we should be able to provide a "map" of sorts showing how the larger community of organizations are related.

This may also provide some indication of the role of Zapatista-related sites in this network, by indicating the "central" domains, those that appear to be closest to the largest number of other domains. A concentration of links to one domain or group provides us with one measure of centrality, but not a complete picture. We can also measure "betweeness," to identify domains or subgroups that are most likely to be passed *through* even if they do not necessarily have a large number of links to them directly. Such observations of the structural position of Web sites have already found a place in the techniques used by search engines like Google that determine the most "influential" site among a group by measuring the hyperlink structure surrounding it (Henzinger 2001). Using a similar approach, we can identify the central domains within this network, and those that act as important intermediaries. We can also undertake a similar analysis using the groupings of tightly knit domains to determine which of these *groups* plays a central role in the network.

Crawling the Zapatista Web

Collecting hyperlink data from the Web raises special challenges. Although a number of approaches have been taken to gather samples, we decided to snowball sites from the EZLN Web site (http://www.ezln.org/), which can be considered the most important public organ of the Zapatista movement. Making use of a custom Web crawler,[1] we collected the first 250 pages of the EZLN site and coded target sites to determine whether they, too, were activist NGOs. For the purpose of this coding, Web sites of "activist NGOs" were defined as those that 1) were clearly noncommercial, 2) were nongovernmental, 3) indicated that they had a particular social mission, and 4) had a significant and obvious "real-life" component or membership that was engaging directly in activism. This would exclude, for example, the Open Directory Project, which contains links to many NGOs but is not an activist organization by this definition.[2] Additionally, in order to disambiguate these sites, we examined only those with a unique domain.

Thus, if an NGO was hosted by a commercial enterprise, or by a university, it would not be included. While these restrictions may have excluded valuable organizations, in practice we found this to be a reasonable approximation of extant NGOs and grassroots groups with a Web presence.

The Web crawler remained within these specified NGO sites, and crawled only the first 250 pages of each domain in turn. Data was collected for activist NGO sites within a radius of two hyperlinks from the EZLN site. Ideally, the "complete" network of activist NGOs would be collected, that is, any activist NGO site that was connected to the EZLN site by any number of hyperlinks. Given that such a collection would require a considerable (indeed, an indefinite) amount of time to crawl and analyze, the present study established the two-link radius as an arbitrary limit. In total, the crawler accessed about one-hundred thousand pages, recording several million hyperlinks, at the end of 2001. We collected these data and sorted them by domain name, for a total network of 392 domains. For the purposes of doing a network analysis, we arranged the hyperlink data among these 392 domains in a square, asymmetric matrix, with elements of the matrix indicating the total number of hyperlinks from each domain to each other domain.

Teasing out the Structure of the Zapatistas' Social Network

As noted earlier, two manipulations of these data are required in order to answer questions about the organization and importance of particular domains to the larger region. First, while examining the interactions of over a thousand domains remains difficult, we can gather many of these domains into cohesive subgroups, and then examine the interaction among those groups. Second, we can look both at the original data and this derived network to determine the importance of local domains or groups to the entire network.

The ways of grouping network data can be roughly divided into those techniques that come from graph theory and those that are applied to clustering non-network data. Cluster analysis—which belongs to the latter category—allows us to more easily make use of the information regarding the number of links between two domains. Since the 392 nodes are already a simplification of a much larger network structure (which includes many pages in each domain), it is important that we retain as much data as possible. When measuring the strength of these ties, a standard hierarchical clustering provides us with a more workable set of groups, especially if it is clear that the clusters generated by an analysis of the hyperlinks are in some way explicable in terms of qualitative groupings.

A total of eighty-three (21 percent) of the domains are peripheral, linked weakly to a single domain that is more central to the network. Just as very few hyperlinks unite this group with the whole, it is difficult to find a common topical thread among these domains. Indeed, several organizations are directly related to the Zapatista movement (e.g., the Comité de solidarité avec les peuples du Chiapas en lutte in Paris), but they exist in this peripheral region because they lack strong hyperlinked connections with more core groupings. Most of these are smaller organizations, more than half with a Latin American focus. There are, however, some anomalies, like Doctors Without Borders, which may appear at the periphery merely as an artifact of the limited sample.

If we cluster the network to the point at which every member has, on average, no less than two links to each other member, and we exclude clusters with less than four domains, we are left with thirteen core groupings, as listed in Table 7.1. Note that the divisions are not as cohesive as the labels might make them out to be. Domains with sites that mainly treat women's rights may be found throughout most of the groupings, and even the Zapatista Information includes the Amazon Watch and Oil Watch Web sites. While a content analysis might not group these domains together, the strong linkage patterns make clear that they are closely related. These groups make up the core 40 percent of the network, and as noted earlier are considerably more interconnected than the other domains crawled.

A second grouping, labeled Zapatista Global Support, is closely linked to the Zapatista Information group, but tends to contain sites that are based outside of Latin America, in languages other than Spanish, and contain more general impressions of the struggle, as opposed to more current news and information found in the "Zapatista Information" group. Most of the remaining groups have been identi-

Table 7.1 · Clusters of Highly Interlinked Domains

Label	# of Domains	Example Domains
Zapatista Information	18	EZLN, FZLN, Laneta, CIEPAC
Zapatista Global Development	23	Accion Zapatista (at the University of Texas), The Irish Mexico Group, Labor Net, Nodo 50
Cultural Exchange	5	International Service for Peace, Global Exchange, Afrocubaweb
Latin American Focus	8	Partners for the Americas, Nationa Council of La Raza, Aspira Organization
Women and Development	11	Sisterhood Is Global Institute, Women Action Network, HIVOS, Synergos
Human Rights	22	Human Rights Watch, Amnesty International, Human Right in China, International Gay and Lesbian Human Rights Commission
Women's Rights	22	Women's World Summit Foundation, Instituto Social y Politico de la Mujer, Equlity Now
Peace Groups	19	Witness for Peace, Center for the Advancement of Non-Violence, Radio for Peace International
Health and Family Planning	7	Fundacion Mexicana para la Planificacion Familiar, GIRE, North American Women's Education Resource Center
Guatemala	4	Rights Action for Central and South America, Guatemalan Human Rights Commission/USA
Grassroots Media	4	Paper Tiger Television, Adbusters
Trade Issues	10	Global Trade Watch, Bretton Woods Project, Central and Eastern Europe Bankwatch Network
Miscellaneous	4	Refuse and Resist, Revolutionary Association of Afghanistan Women

fied by an overarching label that describes the makeup of their constituent domains in broad strokes.[3]

Having reduced the original sample to a more manageable set of groups, we can also reach some understanding of how these groups are interrelated. Figure 7.1 shows the relationships between the thirteen subgroups. Note that although quite distinct from one another, there is a strong relationship between the two directly Zapatista-oriented groups and the largest grouping, that of domains that generally treat human rights issues. An examination of the links surrounding these three large groups provides a more visual depiction of the relationship suggested qualitatively by a number of researchers who have described the Zapatista network. While the sites that are directly related to the Zapatista network may not link the global networks of NGOs together, the secondary tier of Zapatista-related Web sites do perform this bridging function, drawing together disparate social movements. Visually, this can be seen in the strong sets of linkages from the Zapatista Global Support node to other subgroups in the network. Of particular interest is its function as the group most closely linked to the women's rights cluster.

We can gauge this level of centrality more directly by analyzing the linkage structures among the groups. The simplest way to look for centrality is to look for the groups that have the highest "in-degree," or links leading from other subgroups to that subgroup. The Zapatista Global Support is clearly at the lead here, with a total of 339 links from other groups leading to it. The Human Rights subgroup, by comparison, is the second most popular destination, with 227 links. These two groups also produce the largest number of outbound links, 305 and 263 respectively. Of course, we might expect this level of linkage, given simply the size of these subgroups and the domains they contain. If we find the proportion of outdegree to indegree, we are able to see that the Grassroots Media group is the "stickiest," being the target of more than twice as many links as it has outbound, while the Guatemala subgroup neatly reverses this relationship and is the target of half as many hyperlinks as there are links from the subgroup to other destinations. All of these measures compare the subgroups only to their neighbors, without providing an overall picture of the network.

Lin Freeman (1979) described two measures that help to ascertain

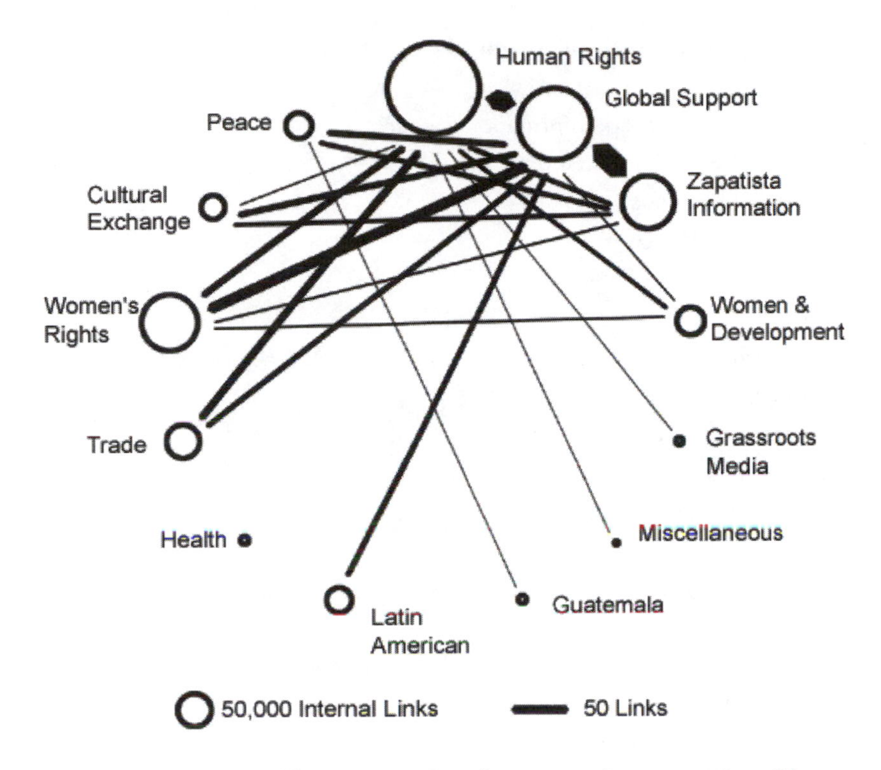

Figure 7.1 · A map of interconnections between subgroups. Ties of fewer than ten hyperlinks are not indicated.

centrality of points to the entire network: "closeness" and "betweeness." The first of these provides an indication for each node of a network of how far it is, roughly, from all of the other points. The second indicates the necessity of passing *through* a given point when moving from one node to another on the network. These provide us with some good indicators of the importance of a given node to the entire network. Unfortunately, they do not take into account the strength of given connections, the number of hyperlinks.[4] Therefore, measuring closeness in this case is facile. Both of the Zapatista subgroups are connected to every other subgroup, leading to the highest possible closeness. When we measure the betweeness of the subgroups, we find that "Zapatista Global Support (17.4, using UCINET; Borgatti, Everett, and Freeman 2001) is the greatest bridge, followed closely by the

Table 7.2 · Domains with Greatest "Closeness" (>1.164)

Latin American Network Information Center
Zapastista Network of Pittsburgh
Amnesty International
Human Rights Watch
CIEPAC
EZLN
Global Exchange
Green Net
Le Neta
Nono 50
School of the Americas Watch

Table 7.3 · Domains with Greatest "Betweeness" (>1200)

Latin American Network Information Center (LANIC)
Zapastista Network of Pittsburgh
Human Rights Watch
Amnesty International
Green Net
Le Neta
EZLN
Association for Progressive Communications
Global Exchange
Nodo 50
CIEPAC

Zapatista Information subgroup (15.8), with peace groups taking the third position (9.9).

More detailed information can be derived by examining the original 392 domains, rather than the network of subgroups. Tables 7.2 and 7.3 list the domains with the highest measures of closeness and betweeness. In the case of sites that are most central to the network (i.e., have a high degree of closeness), it is clear that they aim to reach large audiences, and in some cases do so directly through the Web. Several of these

(LANIC, La Neta, and Nodo 50) provide wide links to related organizations and act as hubs for the larger network. Within the collected network,[5] the EZLN site and other Zapatista-related sites clearly play an important role. When we look at measures of betweeness, we find a very similar list. However, in this case, the LANIC domain is far and away the most important site for connecting the network together; its betweeness proportion is greater than the next ten sites combined. Note also the presence of the Association for Progressive Communications (2001), an organization that aims to "advocate for and facilitate the use of information and communications technologies (ICT) by civil society," an objective born out in their position within this network.

Conclusion

Overall, we find strong support for the widely averred claims that Zapatista-related sites are central to global NGO networks and help to bind them together. Were the Zapatista-related sites removed from the network examined here, the resulting network would consist of a much more balkanized set of Web sites, groups that might link only through the most circuitous of paths, if at all. Setting aside the content of these sites, it is clear that the Zapatista movement has had an impact on the structure of an important region of the Web.

The greatest limitation of the study presented herein is one of scope. While over one-hundred thousand pages may seem like a large number, it does not accurately represent the larger space of the NGO networks. How far would be enough? As we collected this sample, at each step we evaluated activist NGO sites that were linked to the current crawl. This number begins to shrink at about fifteen hundred NGOs; however, the network of hyperlinks becomes increasingly sparse as the number of sites increases. Nonetheless, a much larger sample would represent a useful network not only for the purposes described here, but also to help answer other research questions. Naturally, there is the possibility that NGOs exist that do not link to a single other NGO Web site. Given the sites that have been included in this relatively small crawl, however, it seems that such sites are not plentiful. Certainly, the role of hyperlinks in online discourse among activists merits further investigation.

In a recent interview, Gabriel García Márquez asked Marcos about the place of literature in his life. He replied that as a child he came to think of language "not as a way of communicating but of building something."[6] Many have written about the networked and virtual movements that the Zapatistas epitomize. If we are to discuss these networks as social fact, as something being built through discourse and action, we must do more than acknowledge their presence. We must tease out their structure and make sense of how they are used. Until they are made clear, they remain a part of the sublimated structure of social movements, an ideology rather than a practice.

Notes

1. A Web crawler, sometimes called a "spider" or "robot," collects pages from the Internet by automatically following and recording hyperlinks. It is commonly used, for example, by search engines that are indexing the content of the World Wide Web.

2. They may not agree with this assessment. The page defining the project (http://dmoz.org/about.html) shares a number of rhetorical devices we might associate with an NGO. Nonetheless, when applied to the sites that made up the collected network studied here, these criteria were relatively unambiguous.

3. Two of these require some clarification. The four domains listed under Guatemala seem to be interlinked mainly because of their activities in that country. The Miscellaneous domains appear to be linked only because they share a Web design company.

4. There are certainly measures available to determine centrality based upon the strength of measures (e.g., Bonacich's influence measure), as well as those that more clearly disaggregate indegree and outdegree centrality. Future analysis would certainly benefit from a more extensive analysis of these properties. Given the aims expressed here, and the fact that this may further emphasize already expressed bias (see the following note), we plan to do this only with a larger sample of sites.

5. Naturally, there is a danger here that we are finding precisely what we collected. We would expect the EZLN site to be at the center of a sample that began collection with the EZLN site. Early indications suggest that, not surprisingly, centrality of the EZLN and related sites decrease as a wider net is cast. Nonetheless, this provides some indication of the role of these sites within an arbitrary "distance" of hyperlinks from the EZLN site.

6. "De una o otra forma adquirimos la conciencia del lenguaje como una forma no de comunicarnos sino de construir algo. Como si fuera un placer más

que un deber. Cuando viene la etapa de las catacumbas, frente a los intelectuales burgueses, la palabra no es lo más valorado. Queda relegado a un segundo plano. Es cuando llegamos a las comunidades indígenas, cuando el lenguaje llega como una catapulta" (García Márquez 2001).

References

Adamic, Lada, and Eytan Adar. 2001. "You Are What You Link." Paper presented at the Tenth International World Wide Web Conference, Hong Kong.

Association for Progressive Communications. 2001. "Our Work" Online. http://www.apc.org/english/about/work/index.htm.

Björneborn, Lennart, and Peter Ingwersen. 2001. "Perspectives of Webometrics." *Scientometrics* 50(1): 65–82.

Bonacich, Phillip. 1987. "Power and Centrality: A Family of Measures." *American Journal of Sociology* 92: 1170–82.

Borgatti, Steven, Martin Everett, and Lin Freeman. 2001. UCINET 5.0 Version 5.2.0.5. Natick: Analytic Technologies.

Brunn, Stanley, and Martin Dodge. 2001. "Mapping the 'Worlds' of the World Wide Web: (Re)Structuring Global Commerce through Hyperlinks." *American Behavioral Scientist* 44(10): 1717–39.

Castells, Manuel. 1997. *The Power of Identity* (The Information Age: Economy, Society and Culture Volume II). Cambridge, MA: Blackwell Publishers.

Cleaver, Harry. 1998. "The Zapatista Effect: The Internet and the Rise of an Alternative Political Fabric." *Journal of International Affairs* 5(2): 621–40.

De Angelis, Massimo. 2000. "Globalization, New Internationalism and the Zapatistas." *Capital and Class* 70: 9–34.

Diani, Mario. 1992. "Analyzing Social Movement Networks," pp. 105–35 in *Studying Collective Action*, ed. Mario Diani and Ron Eyerman. London: Sage.

Freeman, Lin. 1979. "Centrality in Social Networks: Conceptual Clarification." *Social Networks* 1: 215–39.

Froehling, Oliver 1997. "The Cyberspace War of Ink and Internet in Chiapas, Mexico." *The Geographical Review* 87: 291–307.

Galaskiewicz, Joseph, and Stanley Wasserman. 1993. "Social Network Analysis: Concepts, Methodology, and Directions for the 1990s." *Sociological Methods and Research* 22(1): 3–22.

García Márquez, Gabriel. 2001. "Habla Marcos." Online. http://www.revistacambio.com/.

Garton, Laura, Caroline Haythornthwaite, and Barry Wellman. 1997. "Studying Online Social Networks." Online. http://www.ascusc.org/jcmc/vol3/issue1/garton.html.

Halavais, Alexander. 2000. "National Borders on the World Wide Web." *New Media and Society* 2(1): 7–28.

Halleck, DeeDee. 1994. "Zapatistas On-line." *NACLA Report on the Americas* 28(2): 30–32.

Henzinger, Monika. 2001. "Hyperlink Analysis for the Web." *IEEE Internet Computing* 5(1): 45–50.

Jackson, Michelle. 1997. "Assessing the Structure of Communication on the World Wide Web." Online. http://www.ascusc.org/jcmc/.

Jones, Quentin, and Sheizaf Rafaeli. 2000. "Time to Split, Virtually: 'Discourse Architecture' and 'Community Building' Create Vibrant Virtual Publics." *EM- Electronic Markets* 10(4): 214–23.

Kim, Hak Joon. 2000. "Motivations for Hyperlinking in Scholarly Electronic Journals: A Qualitative Study." *Journal of the American Society for Information Sciences* 51(10): 887–99.

Kleinberg, Jon. 1999. "Authoritative Sources in a Hyperlinked Environment." *Journal of the ACM* 46(5): 604–32.

Kling, Rob. 2000. "Learning about Information Technologies and Social Change: The Contribution of Social Informatics." *The Information Society* 16(3): 217–32.

Mathews, Jessica. 1997. "Power Shift." *Foreign Affairs* 76(1): 50–66.

Melucci, Alberto. 1996. *Challenging Codes: Collective Action in the Information Age*. Cambridge: Cambridge University Press.

Park, Han Woo. Forthcoming. "What Is Hyperlink Network Analysis? New Method for the Study of Social Structure on the Web." *Connections*.

Rheingold, Howard. [1993] 2000. *The Virtual Community. Homesteading on the Electronic Frontier*. Cambridge, MA: MIT Press.

Ronfeldt, David, John Arquilla, Graham Fuller, and Melissa Fuller. 1998. *The Zapatista "Social Netwar" in Mexico*. Santa Monica, CA: Rand Arroyo Center.

Schulz, Markus. 1998. "Collective Action across Borders: Opportunity Structures, Network Capacities, and Communicative Praxis in the Age of Advanced Globalization." *Sociological Perspectives* 4(3): 597–610.

Scott, John. 2000. *Social Network Analysis: A Handbook*. 2nd ed. London: Sage.

Wellman, Barry. 2001. "Computer Networks as Social Networks." *Science* 239(14): 2031–34.

Wellman, Barry, and S. D. Berkowitz, eds. 1997. *Social Structures: A Network Approach*. Cambridge: Cambridge University Press.

8

Identifying with Information

Citizen Empowerment, the Internet, and the Environmental Anti-Toxins Movement

Wyatt Galusky

Elements of the environmental anti-toxins movement (an assemblage of individuals and groups interested in eliminating or at least abating the pervasive, dangerous effects of toxic chemicals in the environment) have begun to employ the Internet in the promotion of various forms of intervention. Taking advantage of Right-to-Know legislation, database technology, and the World Wide Web, anti-toxins groups have sought to empower individuals to become informed citizens—to know about the toxins in their area, to understand the risks they face, and to change the conditions that allow such risks to proliferate. Visiting the activist Web site www.scorecard.org, the Internet user can find out which industry is emitting what chemical in what area with a few keystrokes and clicks of a mouse. Now people have easy access to information and can use that information to make concerned and conscientious decisions about their community and their lives.

The form of empowerment that this Internet interface offers, however, is neither self-evident nor without cause for concern. As currently constituted, online anti-toxins activism configures the "citizen" as a generic individual in need of expert-mediated information. This composition of the empowered citizen is both limited and limiting. It privileges and even valorizes modes of dependence, making the individual beholden to systems of expertise. It also diminishes the value of the particularity of specific persons with novel and valuable experiences—local experiences not of simple digitally communicated risk but of physically felt toxicity. Lost in this strict digital translation are possibilities of constructing alternate forms of expertise embedded within alternative modes of certification and value. As the anti-toxins movement revisions itself cybernetically, it must take care to pursue actively the kind of world and kind of people it hopes to facilitate. At stake is a robust definition of the citizen that emphasizes the value of local experience and empowers that citizen to have a meaningful voice regarding how these local experiences get constituted.

In Need of Information

The environmental anti-toxins movement, interested in exposing and diminishing the prevalence of toxins in the environment, seems positioned to benefit from the production and distribution capacities of Internet technology. Information, or the lack thereof, has always played a prominent role in the formation of this type of activism and activist. The roots of the modern movement, in the United States, can be found in the tragedies of Love Canal and Woburn, among others. These catastrophes were in part predicated on lack of available information about the presence and impacts of chemicals buried underground and leaking into drinking water.[1] In Love Canal, houses, schools, and playgrounds were built on top of "sealed" and long-forgotten toxic waste dumps (Levine 1982), and in Woburn, community members had unknowingly been drinking water contaminated with trichloroethelyne for years (Brown and Mikkelsen 1990). Not knowing beforehand about the presence of these toxins, people could neither take protective measures, nor could they weigh the risks involved with living in areas subject to exposure to these substances. With the publicity that events like Love Canal received, coupled with the popularity of books cataloging the presence and dangers of potential chemicals (e.g., Rachel Carson's [1962] *Silent Spring*), the problem of toxic waste became a "mass issue," and the environmental anti-toxins movement gained momentum (see Szasz 1994, ch. 3).

In the context of U.S. policy, intervention aimed at ameliorating the effects of a toxic environment has primarily involved the production of information about the presence of toxins in neighborhoods and the promotion of wide-scale availability of that information. This method feeds upon the urgency that permeates perceptions of, and desires to flee, unacceptable toxic risks—the need to know, now, about what toxins may be present. The promotion of intervention as information gathered speed after a rapidly spreading, large-scale international tragedy readily played on public fears about "unknown" environmental toxins. The Union Carbide methyl isocyanate leak in Bhopal, India, in December 1984 (along with a smaller incident at a sister plant in Institute, West Virginia, the next year), gave a boost to movement efforts to create extensive and accessible data (Hadden 1989). In Bhopal, people actually ran toward the gas cloud, and those exposed (and the doctors trying to treat them) were unsure of what kind of gas had been released, and thus could not readily determine what available treatment measures may have been the most effective (see Fortun 2001).

Kim Fortun (2001) provides an insightful analysis of the political and social fallout of the Bhopal tragedy, especially in India and the United States. In discussing the impacts of the incident, she develops the notion of "remembrance" (352). Remembrance involves the process of assigning meaning and drawing insight from an event in the past by remembering that event as one thing, and forgetting that it was also others. Thus, we create normative lessons from the past based upon our visions of what the present is and what the future should be.[2] While blame for the accident in Bhopal circulated between Union Carbide and disgruntled employees, the arguments in the United States focused on the position that the tragedies could have been tempered had community members known about the potential danger and had developed appropriate responses to potential accidents. In U.S. policy circles, Bhopal was remembered as a tragedy involving too little information and emergency planning, and subsequently forgotten as an act of corporate negligence and culpability. Consequently, in 1986, the Emergency Planning and Community Right-to-Know Act (commonly the Right-to-Know, or RTK, Act) was passed. The act mandated the creation of community emergency response plans in the event of the accidental release of recognized toxins, as well as the collection of electronically available data on toxic releases by certain large industries (Hadden 1989). The latter provision created the Toxics Release Inventory (TRI), a computerized database that catalogs the type and amount of chemical released by those industries. Industries are required to supply that data, but such information is self-reported.[3]

With the creation of the TRI, large amounts of computerized raw data on toxic emissions became available—information that could, at least in theory, empower the individual to make informed decisions.[4] The Environmental Protection Agency (EPA) provides access to the TRI on its own Web site, www.epa.gov, in an online database. The EPA, however, explicitly refuses to provide any analysis of the data; the agency will not make any statement about what the various numbers might mean in terms of risk or hazard. The RTK Act provided for the creation of such a database, but not for official declarations on what those numbers say about the risks involved in living in a particular place. Thus, a person living in the 24060 zip code may be able to discover, for example, that sixty-four thousand pounds of toluene were emitted into the air in 1999 (http://www.epa.gov/triexplorer/chemical.htm), but have little idea as to whether such an amount of this chemical substance, from such a plant, in such an area, constitutes an appreciable hazard to her or anyone. Simply put, these numbers on their own mean nothing, outside of some interpretative framework. The U.S. government refuses to provide such a framework, allowing intermediary groups the privilege, and the liability. The information, however, is available to anyone with the means to access it.

"Empowerment" as Access

Offline, the prototypical mode of activist organization that formed around these sets of issues and concerns was the Citizen's Clearinghouse for Hazardous Waste (CCHW), now the Center for Health, Environment, and Justice (CHEJ; see www.chej.org[5]). This organization, founded in 1981 by Lois Gibbs (a legendary anti-toxins activist and resident of Love Canal), has a long history of allowing self-selecting people who become interested on their own to seek help from the CCHW and its resources—monetary, experiential, organizational, and informational. When contacted, CCHW provided help for individuals and groups in their efforts to locate sources of information. The organization also provided guidance for making that information intelligible and politically effective. Last, CCHW looked to foster small, community based groups to take active responsibility for the places in which they live.

The interactive move to cyberspace has been done by Environmental Defense, who endeavors to exploit the digital availability of TRI and the popularity of the Internet, in order to make the information more easily accessible for the individual. In 1998, this nonprofit activist group put www.scorecard.org on the Web. In many respects, the Web site is an electronic version of CCHW, where (connected and) interested parties can find out about toxic emissions near their home, along with what potential steps they can take to remedy unacceptable conditions. Individuals must seek scorecard.org out, then decide whether they are interested enough to try it out. They are then electronically guided to countywide information about emission levels and are provided with some interpretations and analyses as to what those numbers imply regarding the safety of the area and the kinds of risks faced.

In going online, Environmental Defense seeks to take advantage of the Internet as a medium of communication, theoretically increasing availability, exposure, specificity, relevance to the individual (allowing visitors to find the information about the places they are particularly interested in), and ease of distribution. The Web site describes its own uses and benefits as the following:

> Scorecard is the ultimate source for free and easily accessible local environmental information. Simply type in a zip code to learn about environmental issues in your community. Scorecard ranks and compares the pollution situation in areas across the U.S. Scorecard also profiles 6,800 chemicals, making it easy to find out where they are used and how hazardous they are. Using authoritative scientific and government data, Scorecard provides the most up-to-date and extensive collection of environmental information available online. Information is power—once you learn about an environmental prob-

> lem, Scorecard encourages and enables you to take action—you can fax a polluting company, contact your elected representatives, or volunteer with environmental organizations working in your community. (www.scorecard.org/about/about.tcl)

After being invited to "find your community" via the zip code interface, the user is taken to a page that breaks down the "environmental issues" in the corresponding county. On that page is a list of hypertexted pollution categories (air, waste, land, water), along with links to environmental justice information, interactive maps, and intervention strategies. Pursuing any and all categories of interest, the visitor can look up potential health effects and risks involved with various "recognized" or "suspected" toxicants.[6] The information also highlights what is *not* known about various chemicals.[7] Importantly, the Web site provides a risk assessment framework, which translates various toxins into comparable equivalencies by creating Toxic Equivalency Potentials (TEPs) using benzene as a baseline for carcinogens and toluene for noncarcinogens.[8] This framework represents a controversial attempt to present *some* level of risk analysis in the context of these data. Industry experts contend that this comparative measure does not adequately reflect the risk involved and may cause unjustified panic (Foster, Fairley, and Mullin 1998). Although admittedly not perfect, these measures are an important start toward better gauging risks.

What scorecard.org does is function as a digitized interpretative intermediary, providing both a framework and a context in which those emission data become meaningful. It inserts a layer of analysis between TRI and the user, so that individuals can have some idea of what risks may be present in their area. Scorecard.org not only facilitates a means for assessing risk but also constructs the potential visitor as a citizen. The Web site itself is a context that promotes change and environmental protection. This step is important, for knowledge *of* risk does not necessarily carry with it indications of what to do *about* that risk, if anything. In fact, other sites have sprung up to provide risk assessments, but have aimed that information at other possible identities—for example, the consumer. Most notably, the site www.disclosuresource.com (formerly www.e-risk.com) offers information on toxic risk and exposure, but amid rhetoric of protection of economic investment rather than protection of one's community. Knowledge about toxins is packaged as important for ensuring savvy real estate investments and limiting liability. Blurbs on the site include: "*Forget about the lawn. What about the leaking tanks?*" and "Is your dream home surrounded by environmental hazards?" (www.disclosuresource.com/buyerhome_ca.asp?pagestate=California). The user of this commercial site, in purchasing these data, is not expected to share this knowledge with community members, but simply to activate some comparative advantage in knowing, potentially, where not to live. Sites like disclosuresource.com promote avoidance of toxic environments.

Scorecard.org, on the other hand, encourages activism. [9] In packaging risk assessment with intervention mechanisms in the form of faxes and community-group connections, the site promotes the overall message that toxins harm people and should be abated, or at least not made to impact people unequally. Hence individuals should *be informed* about the substances as a prelude to *doing* something about them, to make localities less toxic. Implicitly, scorecard.org presents the data under the assumption that the people accessing it are (or at least should be) interested in staying in one place and fighting for that place—protecting one's home and one's community. The information is meant as a bridge to the surrounding community members in the face of alarming pollution numbers and unacceptable risk. The site is meant to empower the citizen: "Information is power—once you learn about an environmental problem, Scorecard encourages and enables you to take action" (http://www.scorecard.org/about/about.tcl).

The "Informed/Formed-in" Citizen

What kind of citizen, and what kind of empowerment, does scorecard.org have in mind? There is no doubt that the site is an empowerment tool—it provides the visitor with an immense amount of information about chemicals and their potential effects and gives individuals greater access to data about those chemicals with which they share space. It also empowers the movement by accentuating the fact that, in the current toxin-filled world, "To be a citizen is to be a *potential* victim" (Feenberg 1999, 120). Thus, the Web site invites people to consider the risks they face daily and to act with regard to those risks deemed unacceptable. For Environmental Defense, this means of communication is assumed to be a rather unproblematic solver of particular problems—it makes information about toxins widely available, easy to access, and targeted to specific locales. It remains important, however, to acknowledge the extent to which the Internet influences how problems themselves get constituted.

In using computer-mediated communication technology in this manner, scorecard.org focuses on the form of empowerment and intervention overtly suggested by the distributional and interactive capabilities of Internet technology, predicated on access to information. Other forms of empowerment, which highlight increasing political efficacy and inserting local values, become derivative. Empowerment as access erects information as the gateway to effective and meaningful political change. But, is this kind of "informed" empowerment all that empowering? Internet-filtered assumptions about the information's impact in isolation, about its local relevance, and about its overall value to the individual as an empowerment tool must be examined. The relationship the individual has to the information meant to be empowering must be better understood.

A large benefit of Internet technology is not simply access, but rather *easy* access. Cyberspace helps to mitigate the physical space that had limited access prior to this technological intervention. The Internet gives people time to be activists, by making data, expertise, connections, and intervention strategies available all at one place. The would-be activist can become a politically engaged citizen simply by visiting a Web site and clicking a few options. Thus, for those individuals "enabled" by the technology, they can use it to accomplish tasks and gather information more quickly (Sobchack 1996, 80). At scorecard.org, ease is a big selling point, as its most popular feature is its "type in your zip code" option, with the hope "to make the local environment as easy to check on as the local weather" (http://www.scorecard.org/about/about-why.tcl). This easy, wide-scale availability fulfills part of the promise of RTK, in "lowering the cost of information" (Hadden 1989, 5). Now that citizens have easy access to the information, what are they to do with it, and what will it do with them?

Most generally, easy access can imply that substantive sociopolitical change can occur with a few clicks of a mouse. It is important to reflect on how much individuals must still interact with legislative bodies (e.g., the state) in terms of promoting actual political change. Cyberspace does not simply invalidate the state; states adapt to the pressures and possibilities the Internet provides (Deibert 1997) and thus become weaker in some roles but stronger in others (Everard 2000). Going online does not by itself subvert the typical, state-sanctioned technocratic modes of authority and governance that have disenfranchised individuals from input into political decisions (see Feenberg 1999; Fischer 2000). Nor does scorecard.org explicitly offer any alternatives to the realm of technocratic decision-making. Instead, the Web site validates its information as having come from "authoritative" experts, and therefore valuable as such. It does not promote or even acknowledge the value of local, grassroots knowledge production to these types of environmental issues (see Tesh and Williams 1996). Relying so heavily on this scientifically moderated data for its own credibility, the Web site assumes, or at least privileges, a level of disinterested politics based upon irrefutable scientific knowledge—committing users to the technocratic expert/counterexpert contestations and stalemates that typify policymaking (see Fischer 2000). As a consequence, empowered scorecard.org users are still beholden to standards of evidence proffered in policy contexts, where the burden of proof continues to apply to them (see note 1). Putting the information obtained online into use offline, in an activist campaign within one's community, continues to require corroboration, resources, time, and commitment well beyond pointing and clicking. Surfing the Web is much easier than effective political intervention.

On the other hand, scorecard.org does offer different forms of intervention. It calls attention to, and lets the individual target, specific companies that are engaged in polluting. This market-oriented, consumer-driven approach to intervention, where consumers are to alter industry behavior directly through market tactics such as refusing to buy certain goods, appears ideally suited to the design of TRI. If a person does not like the amount of pollution a company is emitting, then that person will refuse to buy products from that company, thereby coercing the company to reduce those emission. Unfortunately, these tactics have the built-in difficulty of being seemingly unable to convey a *particular* message in the context of a market transaction. Beyond this aspect, scorecard.org appears to have more specific problems. In packaging the information into what is important *to* the user, the Web site also packages the user *solely* as an anti-toxins activist, with little local variance. Actual industries situated in actual communities can play a defining role in the make-up of that community, in terms of work and social spaces and livelihood. Thus, form faxes that address the issue generally[10] may not be in the best interests of the more specific needs and goals of members of a locality, especially in a context of hypermedia-enabled flexible, and thus mobile, capital (see Deibert 1997; Harvey 1996).

This empowerment as access to expert-certified information appears to diminish, not enhance, the individual's powers of input into local, particular issues. The user visits the site to learn about the community, but she does not possess this knowledge; she is simply granted access to it. In relating to this expert knowledge, the user of this Web site exists as a simple conduit through which others speak. The people using the data will likely *not* participate in the production of the knowledge that is supposed to be so important to them. Missing knowledge is highlighted by the Web site (see note 7), but it is not suggested or implied that the user's own knowledge is another component in need of being added. The technocratic decision-making process, based on the impersonal contestations of expertise that "vested" interests bring into the equation, is not called into question. A "local" citizen within this process has value only as a proxy for the authority of others (in the case of scorecard.org, this citizen-participant will have entered that particular zip code and followed that particular knowledge pathway). Thus, in confronting industry, the local "informed citizen" does not participate as a possessor of a unique form of expertise, but as a consumer of other experts.

Thus, scorecard.org engages in a kind of deceptive or limited particularness. The site valorizes the specific, local relevance of the information provided, pertinent to the needs of the individual user. But what is local about this kind of access? The information is distilled and repackaged *not* based upon individual, locally situated needs per se, but on more general prescriptions that are redeposited in the lap of the individual. The particularity is geographic, or spatial, not local, or placial. Scorecard.org facilitates access on behalf of "the concerned citizen and potential anti-toxins activist," not particular citizens with their own particular needs and their own insights and experiences to offer. In being "informed" by specific systems of expertise that digitally configure both "community" and "risk," citizens themselves become "formed in"[11] those self-same systems, articulated by others, for purposes designed by others. Place-specific values and communal, distinct experiences do not play into it. Instead, "local information" stands as code for "general heuristic applied within geographic area."

Thus, one must assess the cost of using the Internet to communicate certain messages. Like other media of communication, the Internet impacts both data and users and favors certain forms of interaction and organization over others (Deibert 1997). The Internet changes not only how we interact with the world, but also how we view the world. As Vivian Sobchack (1996) has noted about communication technologies in general, "our relationship to technologies that instrumentally mediate and thus transform our perception and forms of communication is further complicated by the fact that, so transformed, our perception and expression are *both* amplified *and* reduced" (81). In the context of Environmental Defense's use of this medium, preoccupation with empowerment through Internet access reduces the concept of individual-ness as a locally situated and uniquely experienced possessor of information, and it amplifies individual-ness as an Internet-enabled, geographically dispersed user of information. An individual does not contribute local knowledge, but participates in common knowledge.

This contrast between "access" and "possession" or ownership exemplifies the kind of relationship to information fostered by Internet information technologies (see Hayles 1999). As a modality[12] of communication, the Internet privileges "retrievability" over "repeatability" (Feenberg 1995, 136), enabling symbolic forms in the guise of information or images, media or messages, to be available (theoretically) on demand. As Feenberg (1995) notes, information that is retrievable online does not require the immediate presence of another person—the data are in theory permanent and available for access on the command of the individual user. But a lack of immediate presence does not equate to no presence. The information as coded and made available carries with it normative decisions about proper knowledge and appropriate modes of access, decisions made by others. This kind of information retrieval empowers the user through the simple binary choice of use or not.

Of course, it has been argued that people need not, and indeed do not, take information at face value. They will critically engage the data and make more use of them as they see fit. Beck (1992) notes that, in this current climate of competing knowledge claims, "science" loses its unified voice—certified experts challenge the claims of other, equally certified experts. To this end, people *cannot* simply act as passive receptacles of expertise, but have to construct actively themselves. This cacophony of disparate expert declarations creates a vacuum of certified authority, empowering the individual to decide pertinent political and moral questions. As Giddens (1992) has proposed, access to more information, to the bevy of contradictions and contrary expert pronouncements on everything from diet to death, forces people to craft their own expertise for themselves.

On this view, scorecard.org does not contain expertise as much as it simply enables it. People are empowered not by their choice to use this information or not, but rather to use this information from among several other forms (e.g., as a real estate investor, a concerned citizen, or both). In addition to being contrary to the stated purpose of the site, this train of thought also begs the question as to whether this state of affairs would prompt a reevaluation of the basic constitution of expertise. The individual as expert, in this context, is an adjudicator between competing claims, cobbling together a hodgepodge of pronouncements on a variety of issues that he was told to be concerned about. There need not be any underlying, integrative logic. The integration of these expertises, locally situated within an embodied self, is not necessarily secured in a singular identity of expert. Internet technology can enable individuals to participate in multiple, virtual identities (Turkle 1995). This production of *expertise* is expressed as a consumption of *expertises* (Luke

1989); the notion of expertise is not itself transformed or undermined. No new, previously disempowered voice is necessarily added.

Scorecard.org has provided a service in making previously difficult-to-obtain or nonexistent "expert" information widely and easily available. In so doing, however, it has also reinforced a system of expert-mediated dependence, in which certified information predicates political validity and has weakened more systematic critiques that challenge technocratic decision-making processes and exclusive definitions of expertise. The level of empowerment scorecard.org offers is predicated on even more *loss* of power on alternate sides. In this case, people are not asked to formulate their own knowledge, but to seek out "official" knowledge from certified sources (or at best have their own experiences sanctioned by more legitimate experts). Using the polluters' own information against them has some appeal. The notion of reliance is not subverted, however, but simply reestablished through empowerment as access. People are told to take the information and make of it what they will; they are not asked to formulate what knowledge will be important and why. They are not asked to define the parameters of their own lives in ways that may not be organized around a yearly tally of emissions. Environmental anti-toxins cybercitizens become empowered (and formed) within systems of dependence.

Reconsidering Empowerment: From "Informed" to "Willful"

The point so far has not been to argue that citizen empowerment is not important to the anti-toxins movement, nor that such a quality is impossible on the Internet. People need to make informed choices about the world in which they live—to wisely weigh and adjudicate risks that are not going to go away (Beck 1992). Empowerment, however, needs to be revisioned to have a particular emphasis. Empowerment can imply actively subverting existing levels of expertise, taking ownership of particular problems, gaining access to the creation of knowledge (or at least acknowledgement of the value of local experience), and having an investment in making local knowledge a meaningful policy player. It may be that escaping the dependence on the technocratic systems of decision-making is just what the anti-toxins movement should be focusing on.

In discussing the roles of chronically ill patients in clinical trials, Andrew Feenberg (1995, 117) suggests, following Hans Jonas, that there is a qualitative distinction between giving consent or permission to be involved in an experiment and willfully participating in that experiment. The consensual patient, even when "informed," still plays the part of the "proverbial 'guinea pig'" (Feenberg 1995, 117); the patient's fate is still in the hands of others, based upon values and options "expertly" configured. The willful patient, on the other hand, provides much more input into the experiment and collaboration with the researchers, in choosing what results would constitute success and under what conditions risks might become "acceptable." Feenberg extracts a larger lesson from these versions of a technological system (experimental medicine)—about the ability to change technological systems from within, coupled with the need for greater input in the design, not just implementation, of those systems. With the world itself now a roughly crafted experiment testing the long-term effects of toxins on humans and their environment (Beck 1992), people need to advocate for more willful participation in the parameters and goals of that experiment. Making a transition from ignorant pawn to informed, consensual subject through the intervention of available expert information, however, may not represent a big enough improvement. How the world, or at least "community," gets designed and constituted is a question that more people deserve to ask and answer.

For environmental anti-toxins activists, there are severe limitations involved in being beholden to any kind of traditional expert information and mediation. In remembering the history of the movement, it is important not to forget that citizens became activists not simply to demand information, but also to demand the right to speak for themselves. The anti-toxins activist identity has been mobilized around both a lack of information about hazards faced, and the systematic denigration of local forms of knowledge and experience. In the case of Love Canal, the residents ran into two frustrations—that experts (government officials, city planners, etc.) never provided them with information about the toxins they lived with or their potential effects and that experts subsequently denigrated the validity of their experiences of harm and of the data they collected (Gibbs 1982). Community members were not entitled to speak on their own behalf, whether on the value of their area or on the reality of their experiences. Even sympathetic scientists were denied their typical authoritative place in matters of policy (Levine 1982). In the face of these experiences, forms of activism need to call into question the right of experts to have such exclusive privilege and to demand an accounting of credible knowledge.

Empowerment can also be considered to be having the right to speak for oneself, not just decide for oneself. Interventions like these invoke instances of the so-called not-in-my-backyard (NIMBY) phenomenon, which exists as a rather pejorative label for promising interventions of community members on behalf of their community. For example, attempts to site industrial waste facilities or landfills, according to Frank Fischer (2000), can be considered in a class of "'wicked' problems [that] lend themselves to no unambiguous or conclusive *formulations* and thus have no clear-cut criteria by which their resolution can be judged" (128; italics mine). The issue is not the solution to the problem, but the construction of the problem itself. Thus conflicts about the nearness of toxins are not the result of knee-jerk reactionary irrationalism, but instead are contestations made by empowered citizens about the value of communities and the role of expertise in everyday life. Instead, Fischer supports a notion of empowerment that makes the term "expert" much more egalitarian. For Fischer, increased citizen involvement in almost all areas of policy may not only be more effective, but even necessary. In contrast to the technical rationality of science, he elaborates another form of rationality—cultural rationality—which centers around "personal and familiar experiences rather than depersonalized technical calculations" (Fischer 2000, 132). He advocates the importance of scientific knowledge collaborating with local forms of knowledge, rather than displacing them (194). This collaboration still operates under the auspices of "science," but the science as practiced does become more open to the input of others. Expertise is refashioned to be equally important locally as scientifically.

The value of this type of intervention, especially in increasing the effectiveness of activism, is furthered by Shemtov (1999). He promotes the utility of an "ownership" frame for understanding how grassroots environmental justice groups expand their goals and objectives beyond single-issue organizing and become more future-looking. The frame includes three characteristics: diagnosis, prognosis, and motivation. This latter quality illustrates the importance of groups feeling a sense of entitlement, exclusivity, and expertise in speaking about an issue either through their geographic particularity or issue-related experience. Ultimately, efforts to disseminate new and valuable information should not usurp potential ownership of the local problems in which people might be willing to carry the issue forward. People must then be able to use the information to springboard into more community-oriented forms of research and knowledge production.

Both Fischer and Shemtov remain invested in the authority and usefulness of "science," and whether science can facilitate radical changes in the current power structure is not obvious. Within the current confines, however, both advocate for a more open definition of science if nothing else. In this age of risk, these demands become more and more credible and important. Beck (1992) argues persuasively that amid the current contestations of expertise and counterexpertise vying for epistemological sanction, a situation emerges "of great ambivalence. It contains the opportunity to emancipate social practice *from* science *through* science; on the other hand, it *immunizes* socially prevailing ideologies and interested standpoints against enlightened scientific claims, and throws the door open to a feudalization of scientific knowledge practice through economic and political interests and 'new dogmas'" (157). These alternate possibilities, on the one hand wresting politically viable knowledge claims from the exclusive hands of technocratic scientific expertise, and on the other explicitly entrenching science as just another tool of the powerful, make activist interventions aimed at the former all the more important. With the danger of science becoming purely "performative" (Lyotard 1984), not seeking any pretense of "truth" but simply performing a service for those in charge while still occupying its decision-making role in a technocratically dominated political sphere, critiques that challenge the persuasiveness of expertise and seek to revision certified knowledge production are vital, for activists of all stripes.

Conclusion

Any group that engages an Internet interface must take pains to evaluate the assumptions embedded within that interface, and what sort of empowerment is really afforded the people interested in accessing the site. Is it the "informed citizen formed-in" various modes of previously constituted expert pronouncements, or is it real people taking real account and real charge of their organization and goals based upon the local needs and characteristics of the community and the area?

With the current Internet schematic employed by scorecard.org, many of the important questions are already decided, which hinders the flexibility and control over empowerment endowed within the user. Experts have decided what toxins are important, what counts as an indicator of toxicity, and how to determine a toxic substance in a community. The "cybernetic" vision of environmental anti-toxins activism, this amplification of our perception of the problem to be one of information-poverty, hinders efforts to problematize all forms of expert mediation and to escape expertise/counterexpertise stalemates and ties perceptions of the Internet to hierarchical dispensations rather than increased democratization. In this case, the coproduction of Internet technology and environmental anti-toxins activism tends more toward a form of consumptive empowerment, where the Internet becomes a speedier delivery service of the same old inequalities, in which "informed" activists are empowered to partake in a kind of emissions market. All the while, the expert-mediated digitization of toxins and risk denigrate real confrontations, by valorizing virtual experiences of toxicity over local, particular ones.

Desire for local control and local empowerment, however, can only be met with opportunities to participate at *all* levels of political action, especially knowledge production. Rather than endorsing the authority of expert systems to deal with problems perpetuated by technocratic decision-making, environmental anti-toxins activists should promote digital contestations over the production processes and use of information, through defining "alternative networks" (Luke 1989, 257) of knowledge as more local, and less common. Larger questions remain, regarding whether the Internet can facilitate this process of escaping technocratic modes of decision-making and expert pronouncements of acceptable risk. Can the Internet provide these alternative networks of power through the productive activities of marginal groups and identities (for cautious optimism, see Feenberg 1999; Luke 1989)? Or will the digital age be marked by a new form of class inequality, distinguished by the "interacting" and the "interacted" (Castells 2000, 402)? While these question are still open (and will never fully close), it remains important to engage critically the technology in an effort not to be unduly influenced by it.

Notes

1. Of course, these tragedies were more than the simple result of people not knowing. The willful disposal of potentially toxic chemicals was a regular, and largely unregulated, practice (see Colten and Skinner 1996). Companies and governments operated under the assumption that such toxin disposal practices were fine until someone proved otherwise, leaving the burden of proof on the potential exposee that the toxins are harmful, rather than on the exposer to show that introducing these toxins into the world is "safe." As Beck (1995, 92) has noted, toxins seem to be in need of help—"please protect the toxins from the people who are threatening them." As noted later in the argument, the government assists now in making some information on toxin release more readily available, but it has not removed the burden of proof from the individual.

2. The act of remembering is a reflexive, continual process. As such, the past can "change" depending upon present experiences, while continuing to influence present choices and future possibilities.

3. To understand some of the limitations of the data: The number of recognized toxins now stands at about 650, and industries can petition to not make their emissions public if they can show that such information may lead to a breech in national security, in that the site could become a terrorist threat based on the chemicals present (Hadden 1994). That is a disquieting thought on many levels.

4. This kind of citizen empowerment through data collection and availability has been embraced by U.S. regulatory agencies (and to an extent by industry). Such measures forego costly and cumbersome legislative oversight on the actions of industry, on the assumption that informed citizens will do their own regulating through making market-based decisions based upon their engagement with that information. In Europe, similar legislation has been passed regarding the development of plans in case of toxic release. The Seveso directive, however, is conceptualized more as a "need to know," rather than a right to know, with governments taking a more active role in both defining prob-

lems and providing citizens with readily understandable recommendations (see van Eijndhoven 1994, 127).

5. While CHEJ does have a Web presence, the function of the site is to advertise the organization and its many programs, but not to provide any particular data about toxicity. The group still carries out its interventions using older, offline modes of communication and interaction.

6. Scorecard.org breaks down the toxins into recognized and suspected, relying on California's Proposition 65 ("known to the State of California" based on a review of "neutral scientific and regulatory experts") for the former, and on their own analysis for the latter (www.scorecard.org/health-effects/gen/hazid.html). They lobby for the burden of proof being negative and not positive—that is, that the chemical manufacturer must establish that a substance is *not* toxic, rather than people having to prove that it is (see notes 1, 5).

7. According to the site, "The biggest enemy to thoughtful policy on pollution control is ignorance. Gaps in the scientific understanding of which chemicals are toxic, what (and how severe) their health effects might be, and how and where people are being exposed to them will continue to block our ability to regulate pollution properly, reduce emissions as necessary, and deliver the protections that U.S. citizens were long since promised by law, until these gaps are filled" (http://www.scorecard.org/about/about-why.tcl).

8. For carcinogens, TEP = [Added Cancer Risk/Unit Release of Chemical X]/[Added Cancer Risk/Unit Release of Benzene]; for noncarcinogens, TEP = [Hazard Index/Unit Release of Chemical X]/[Hazard Index/Unit Release of Toluene] (http://www.scorecard.org/env-releases/def/tep_caltox.html).

9. Interestingly, one reviewer of both sites failed to make any real distinction, claiming both to be useful for the protection of property value (see Tuttle 2000).

10. For example, see Montgomery County, Virginia, at http://www.scorecard.org/env-releases/county-facilities-faxable.tcl?fips_county_code=51121.

11. Thanks to Timothy W. Luke for suggesting this turn of phrase.

12. This concept is borrowed from John Thompson (1990), who offers a framework for conceptualizing how different modalities of communication transmit symbolic forms (information) from producer to receiver. While Thompson's analysis preceded the explosion of the Internet in terms of widescale use (focusing primarily on television), his means of analysis are adaptable to the current medium and useful for distinguishing the relevant factors. He distinguishes three important variables for considering modes of transmission: the technological medium, which can vary the extent of fixity, reproducibility, and participation that it presumes; the institutional apparatus within which those media are deployed; and the space-time distancing that the modality allows.

References

Beck, Ulrich. 1992. *Risk Society: Towards a New Modernity*. Trans. Mark Ritter. London: Sage Publications.

——. 1995. *Ecological Enlightenment: Essays on the Politics of the Risk Society*. Trans. Mark Ritter. Atlantic Highlands, NJ: Humanities Press.

Brown, Phil, and Edwin Mikkelsen. 1990. *No Safe Place: Toxic Waste, Leukemia, and Community Action*. Berkeley: University of California Press.

Carson, Rachel. 1962. *Silent Spring*. Boston: Houghton Mifflin Company.

Castells, Manuel. 2000. *The Rise of the Network Society*. 2nd ed. London: Blackwell.

Colten, Craig E., and Peter N. Skinner. 1996. *The Road to Love Canal: Managing Industrial Waste before EPA*. Austin: University of Texas Press.

Deibert, Ronald J. 1997. *Printing, Parchment, and Hypermedia: Communication in World Order Transformation*. New York: Columbia University Press.

Everard, Jerry. 2000. *Virtual States: The Internet and the Boundaries of the Nation-State*. London: Routledge.

Feenberg, Andrew. 1995. *Alternative Modernity: The Technical Turn in Philosophy and Social Theory*. Berkeley: University of California Press.

——. 1999. *Questioning Technology*. London: Routledge.

Fischer, Frank. 2000. *Citizens, Experts, and the Environment: The Politics of Local Knowledge*. Durham, NC: Duke University Press.

Fortun, Kim. 2001. *Advocacy after Bhopal: Environmentalism, Disaster, New Global Orders*. Chicago: University of Chicago Press.

Foster, Andrea, Peter Fairley, and Rick Mullin. 1998. "Scorecard Hits Home: Web Site Confirms Internet's Reach." *Chemical Week* 160 (21): 24.

Gibbs, Lois. 1982. *Love Canal: My Story*. Albany: SUNY Press.

Giddens, Anthony. 1992. *Modernity and Self-Identity: Self and Society in the Late Modern Age*. Stanford, CA: Stanford University Press.

Hadden, Susan. 1989. *A Citizen's Right to Know: Risk Communication and Public Policy*. Boulder, CO: Westview Press.

——. 1994. "Citizen Participation in Environmental Policy Making," pp. 91–112 in *Learning from Disaster: Risk Management after Bhopal*, ed. Sheila Jasanoff. Philadelphia: University of Pennsylvania Press.

Harvey, David. 1996. *Justice, Nature, and the Geography of Difference*. Oxford: Blackwell Publishers.

Hayles, N. Katherine. 1999. *How We Became Posthuman: Virtual Bodies in Cybernetics, Literature, and Informatics*. Chicago: University of Chicago Press.

Levine, Adeline. 1982. *Love Canal: Science, Politics, and People*. Lexington, MA: Lexington Books.

Luke, Timothy W. 1989. *Screens of Power: Ideology, Domination, and Resistance in Informational Society*. Urbana: University of Illinois Press.

Lyotard, Jean François. 1984. *The Postmodern Condition: A Report on Knowledge*. Trans. Geoff Benington and Brian Massumi. Minneapolis: University of Minnesota Press.

Shemtov, Ronit. 1999. "Taking Ownership of Environmental Problems: How Local Nimby Groups Expand Their Goals." *Mobilization: An International Journal* 4 (1): 91–106.

Sobchack, Vivian. 1996. "Democratic Franchise and the Electronic Frontier," pp. 77–89 in *Cybercultures: Culture and Politics on the Information Superhighway*, ed. Ziguddin Sardar and Jerome R. Ravetz. New York: New York University Press.

Szasz, Andrew. 1994. *Ecopopulism: Toxic Waste and the Movement for Environmental Justice*. Minneapolis: University of Minnesota Press.

Tesh, Sylvia, and Bruce Williams. 1996. "Identity Politics, Disinterested Politics, and Environmental Justice." *Polity* 18 (3): 285–305.

Thompson, John B. 1990. *Ideology and Modern Culture: Critical Social Theory in the Era of Mass Communication*. Palo Alto, CA: Stanford University Press.

Turkle, Sherry. 1995. *Life on the Screen: Identity in the Age of the Internet*. New York: Simon and Schuster.

Tuttle, Larry. 2000, May. "When You Don't Know an Erin Brokovich." *Christian Science Monitor* 22: 9.

van Eijndhoven, Josee. 1994. "Disaster Prevention in Europe," pp. 113–32 in *Learning from Disaster: Risk Management after Bhopal*, ed. Sheila Jasanoff. Philadelphia: University of Pennsylvania Press.

PART III

Cautionary Readings of Community, Empowerment, and Capitalism Online

9

Wiring Human Rights Activism

Amnesty International and the Challenges of Information and Communication Technologies

Joanne Lebert

Since its inception in 1961, Amnesty International (AI), an international human rights organization, has experienced incredible technological change: from pen and paper, Gestetner machines, and conventional mail to short text messaging, satellite news feeds, and "Web-casts" all transmitted in real time.[1] However, the adoption of new forms of communication has not necessarily led to the abandonment of more established tools, nor is the adoption of new technologies always a desirable development. Internationally, many AI supporters continue to handwrite appeals and mail these by regular post; telexes and telegrams continue to be used in some countries; and while text messaging may be a popular new medium alerting subscribers to "calls to action," cyberactivists are still urged to follow up all e-mailed appeals or electronic petitions with an old-fashioned, personalized "snail-mailed" or faxed letter. Although speed of action is an important strength of information and communication technologies (ICTs), the more varied the medium and means of communication, the more accessible the campaign to both members and potential activists. Moreover, a diverse toolkit allows a greater degree of flexibility when the aim is to influence states and nonstate actors that have widely divergent technological capacities and equally varied views of ICTs.

ICTs, therefore, have come to be viewed as a leading strategic tactic for Amnesty—something that can no longer be conceived of simply as background infrastructure. E-mail, in particular, has had a profound and largely beneficial impact at all levels of the organization. However, ICTs are not a panacea; they have their share of limitations and, in Amnesty's experience, are best used in conjunction with other more traditional communication tools, such as the fax machine and regular mail.

This chapter seeks to explore AI's relationship to information and communication technologies. Two interrelated branches of Amnesty's work are examined. First, this chapter looks at information production and dissemination, the sound, thorough, and highly respected bedrock of Amnesty's reputation and legitimacy. Next, the chapter reviews communication, coordination, and mobilization, which constitute the nucleus of action at the grassroots level. Subsequently, issues of access and representation with regard to online activism are considered. Finally, the implications for and challenges to AI's organizational and managerial culture will be discussed.

Although Amnesty is an international movement, the views presented here are largely based on the experiences of the International Secretariat (or IS)—Amnesty's international "head office" based in London—and on those of the Canadian English-speaking section.[2] These have also been heavily influenced by my own experiences as an Urgent Action Coordinator and as a member of AI, as well as through conversations with colleagues and AI supporters internationally.

Information Production and Dissemination

Research has always been the backbone of AI. Not only does the organization pride itself on the reliability and relevance of its research, but it also strives to make this information freely accessible to human rights researchers and advocates everywhere. The advent of the Internet, has greatly facilitated and accelerated the production and dissemination of this information.

At the grassroots level, researchers are using e-mail to establish, nurture, and maintain their regional networks of trusted contacts.

These contacts, consisting of local human rights defenders and Amnesty sympathizers, can instantly communicate the details of local developments to researchers based in London. Depending on the technology available to them, local activists may even relay photos and other scanned evidence to the IS.

Although the security of these informants is clearly a concern, it may be no greater an issue, depending on the degree of electronic monitoring in-country, than when local defenders use more traditional forms of communication. In other words, there is an implicit risk involved for those who commit themselves to defending human rights in countries hostile to such activities.

The electronic medium, however, can lend added protection to local human rights defenders. When contacts or informants are harassed, the IS may be immediately contacted and prompted to issue and distribute a statement and call to action within hours of the act, or even in anticipation of an impending threat. For example, in June 2001, CALDH (Centro de Acción Legal en Derechos Humanos), a Guatemalan human rights organization, informed Amnesty that it was planning to launch a lawsuit against officials of the former military government of General Ríos Montt for the massacre of over twelve hundred indigenous people in thirteen separate incidents in the early 1980s. Because of the advance notice, AI was able to prepare, and circulate by e-mail, an embargoed statement and accompanying urgent letter-writing action in anticipation of threats to the security of CALDH staff. Upon CALDH's news release, sections were given the electronic "green light" from IS to release Amnesty's press release and action. Within hours, members from around the world began to flood the offices of Guatemalan authorities with appeal letters urging them to guarantee the security of all those involved in the lawsuit.

ICTs have also greatly improved Amnesty's ability to collect regional data. Over the years, countless AI volunteers and staff have manually clipped, sorted, filed, and distributed news items collected from a host of international sources. Researchers were often frustrated by a lack of access to current, country-specific information. Today, much to the contrary, Amnesty is literally swamped by volumes and volumes of data.

The IS has also recently invested in infrastructure to collect electronic news feeds via satellite, which it sorts and distributes—over three thousand news stories per day—to IS staff workstations and a systems database, all in real time. This database is in the process of being standardized internationally. Once sections develop the capacity to adopt and support such a system, it is expected that they too will have access to these news feeds. Moreover, other resource- and campaign-specific databases are increasingly accessible. Sections that have the ability to support such databases have immediate access to the latest campaigning materials, urgent letter-writing actions, AI statements, press releases, member and donor information, and AI images and photos, all of which can be printed and used for campaigning and mobilizing at the grassroots level. Sections may also contribute to the information pool and share their own, locally produced materials with their international colleagues.

The acquisition and circulation of printed reports continue to be difficult and even dangerous for some human rights advocates. Clearly, one of the greatest advantages of the Web and e-mail is the ability to distribute information to these individuals and to otherwise closed or isolated communities more cheaply and with less risk to the user.

In many countries, including China, Vietnam, and Tunisia, Internet traffic is heavily monitored through state-controlled firewalls and Internet service providers (ISPs), which prevent access to sites deemed to be offensive, including those of human rights organizations. However, no security system is foolproof. Regardless of which filter is being used or how, avid users of the Internet always seem eager both to learn how to circumvent imposed constraints and to convey these system failings to fellow users. Some activists have resorted to using proxy servers and other circumvention methods to get around state security. Uncensored Web sites may also be used as portals and gateways to the greater virtual world. Therefore, the threat of state monitors may not be insurmountable, at least not for any significant length of time. Nonetheless, state surveillance remains a central and absorbing concern to any discussion of online human rights activism.

Generally speaking, both the Web and e-mail offer the public and AI members a wealth of human rights–related information. E-mail subscribers can receive electronic notification of developments in the Amnesty movement, at the local, national, and international levels. Once visitors are drawn to a section or IS Web site, they can access an entire library of documentation. Currently, they can search the IS site by country or by theme to find relevant letter-writing actions, reports, news items, or press releases, which can be printed out free of charge.

The challenge for Amnesty has been to avoid turning its Web sites into sinkholes of information—where lengthy documents are stored and essentially forgotten. Archived documents must be interconnected in a relevant, consistent, and up-to-date manner. These must be easily called up and searched by site users, whether they be the media looking for a quick statement or position paper, refugee lawyers searching for evidence to support a case, a local activist wanting a printout for an upcoming fundraising event, or a group of students working on a school project. In other words, AI materials must not only be available, but also must be easily accessible to a wide range of users from human rights professionals to lay persons. Both the IS and AI sections recognize that Amnesty's content must be set within an interactive framework that anticipates user needs. "One-stop surfers" should be offered personally tailored means to deepen their commitment to, and understanding of, human rights.

Unprecedented and unfettered international access to AI content raises yet another significant challenge to AI documentalists and strategists. To what extent should Amnesty's work be available electronically, and more importantly, to whom should it be made available? Different levels of commitment and membership to AI require differing levels of access to Amnesty content. For example, country coordinators—local volunteers who provide regional expertise to sections—require access to electronic information pertaining to their geographic area of interest. The general public, AI members, and other volunteers may be restricted from viewing this same information because of its sensitivity. Furthermore, it is practically impossible for Amnesty to verify the identity or motives of an individual or member who visits its sites. How does AI ensure the proper handling of reports, documents, and campaign materials once a user has downloaded them? How does Amnesty restrict access to some and yet maintain a reliable distribution pattern so that others can be encouraged to use its resources? The risk of manipulation, misinformation, and misrepresentation is very real, and all have, at some point, tarnished the legitimacy of the organization.

Amnesty is actively working toward minimizing these risks. Among other strategies being considered, passwords may be introduced as a means to differentiate varying levels of membership and access. Also, more than ever, staff members are being urged to consider the intended audience, the medium of dissemination, and the appropriate security level of a document—confidential, members only, or public—prior to its production.

As ICTs continue to facilitate both communication and public access to AI, the organization has been forced constantly to review how to protect itself from abuse and defamation. Paradoxically, greater public interaction seems to have been met with an entrenchment of boundaries, creating lines that delimit "us," human rights defenders, from "them," those unfriendly to the cause.

Communication, Coordination, and Mobilization of Action

In terms of communication, coordination, and mobilization, e-mail has had a profound and largely beneficial impact on the internal workings of AI—more so than any other communications tool in the organization's history. Once access to e-mail is acquired, it provides a convenient and inexpensive alternative to the telephone, fax, and conventional mail. Contrary to a fax machine, e-mail can be used to contact members of a large mailing list simultaneously and reliably, without the cost of paper, which remains exorbitant in many countries. And, because speed of action is crucial to preventing and stopping human rights abuses, e-mail, although it does have its limitations, has lent itself quite readily to activism.

One of Amnesty's program areas, the Urgent Action Network (UAN), has particularly benefited from this medium. Amnesty's seventy-five active UANs comprise a web of letter-writing members who respond to urgent cases of human rights violations by firing off letters of appeals to relevant authorities, often within hours of having received their call to action. The results can be impressive. In one-fifth of the 499 UA cases in the year 2000, Amnesty was able to document positive developments: Torture and/or death threats ceased, the "disappeared" were found to be alive, investigations into violations of human rights were initiated, medical attention was given to sick prisoners, death sentences were commuted, human rights defenders were protected, or the rights of asylum-seekers were respected.[3]

Although Amnesty cannot prove that its letter-writing actions directly or solely influenced these positive outcomes, anecdotal evidence from former prisoners of conscience and victims of human rights violations, in addition to statements made by lawyers and government authorities, suggests that these appeal letters have a powerful impact. While a direct positive correlation between e-mail usage and positive developments in UA cases has yet to be proven, e-mail has contributed to this form of activism by significantly increasing the speed and scope of action. It may be that the lack of clear results can be explained by the fact that many UANs or members are just now switching over from conventional mail to e-mail.

In sections where e-mail has been widely used, as in Canada, the UAN has clearly felt its impact. Upon investigation and confirmation of a human rights violation—one that requires immediate and mass mobilization—researchers draft a UA, which is distributed, usually electronically, to UAN coordinators around the world so that they, in turn, may render these to their respective members. Because well over 80 percent of Canada's English-speaking section members' receive their UAs via e-mail, letter-writing can begin literally within hours of the IS having received word of a human rights violation.

Speed of action also lends itself to accuracy. As soon as a UA is issued, the facts of the case (such as a person's place of detention or state of health) are all subject to change, particularly when the authorities become aware of international concern. When facts have changed and members are not informed in time, authorities can dismiss their letters as inaccurate and their concern unfounded. Correct, up-to-date, detailed information presented by individuals and groups around the world tends to unnerve and sway authorities. The faster facts are provided, the more likely they are to be accurate. Moreover, as an added benefit, accurate information used in a timely fashion contributes directly to AI's reputation as a reliable source and effective strategist.

UAN members also recognize the importance of speed of action in their human rights work. As a result, those who receive their UAs via e-mail are often frustrated by AI's reluctance to provide them with the electronic addresses of relevant authorities. While UA cases may be drafted and distributed within a number of hours, members are generally forced to use conventional mail—or fax, for those who can absorb the costs of transmission—to deliver their appeals to the appropriate authorities. The use of electronic mail in this final step of letter-writing action—where the AI member contacts the authorities in question—is subject to some controversy within Amnesty.

Many question the effectiveness of e-mail when it comes to influencing abusive governments and nonstate actors. The problem stems from the fact that, to date, AI really does not know, in any great detail, how authorities respond to electronic messages. They may only read the subject line and delete the message. They may simply shut down their e-mail accounts if they are flooded with appeals. Conversely, countries with unreliable telecommunication systems may not be able to support any serious and sudden influx of e-mail or faxes. Also, governments may be suspicious of those who send appeal letters, as their origin may be obscure. They may even suspect that a single person is responsible for a multiplicity of e-mails. Moreover, the tendency to use an informal tone in electronic messages and the playful addresses used by some (e.g., satansmonkey@ or goodtimegirl@) may be deemed offensive and may even have a counterproductive effect in the end.

Perhaps most importantly, e-mail may simply not have the physical weight and presence of a hardcopy letter delivered either by fax or conventional mail. Amnesty members have often exercised their ability to fill authorities' offices with bags of mail—with every letter physically demanding action on behalf of an individual. Finally, most governments have a responsibility to sort, document, and file these bits of correspondence, which may make dealing with hardcopy letters a comparatively more onerous task.

Generally, it has been Amnesty's experience that e-mail, for all its positive attributes, is most effective as an activist's tool when used in combination with more traditional methods such as the fax machine or conventional mail. Again, its impact has been real, profound, and extensive—more so than any other communications tool in the organization's history. However, in AI's case, its contribution has remained largely limited to the areas of communication, coordination, and mobilization of activists—areas that are absolutely crucial to the successful workings of any grassroots movement. The activists' ultimate contact with and influence over those in power, however, continue to be carried out by more traditional means, such as hardcopy correspondence, demonstrations, and face-to-face contact, all of which are emotionally charged and involve physical presence.

With regard to its communication and mobilization role, e-mail has been particularly useful to AI activists in large and geographically dispersed countries. In Canada, for example, members had long complained about feeling isolated from other members. They had complained of a lack of support and direction stemming from lack of access to key staff. E-mail has allowed Canadian members of AI, among others, to work more closely together in-country and transnationally. Members who share a particular thematic or geographic interest have been able to network, brainstorm, consult, plan, and mobilize without having to incur the costs and inconveniences of long-distance travel. Consequently, events can be organized far more quickly (provided such an online discussion is well mediated) and actions can be broadcast across program areas more easily so that local activism is better coordinated and supported. For example, if a local group is staffing an Amnesty information booth at a particular event in their community, the UAN coordinators may e-mail the activists the most recently issued UAs for passers-by to act on. At the Gay Pride parade, for example, the UAs on display will pertain to lesbian, gay, bisexual, or transgendered individuals who are currently at immediate risk of danger because of their sexuality.

Amnesty staff members have generally become more accessible to the membership via e-mail and are thus better able to respond to requests for support. However, as much as the electronic medium has facilitated communication among human rights activists, the need for periodic face-to-face contact has not been removed. Virtual discussion groups and forums may be held together by a common interest, but if they are to last in any meaningful way, they will greatly benefit from face-to-face interaction.

The Web's impact on the internal workings of AI has been far less dramatic than that of e-mail. Without a doubt, the IS, section, and local Web sites have been valuable sources of information for members. However, the Web has also proven to be a source of frustration or disappointment for some. Local group members, for example, who have created their own personal AI Web sites, occasionally speak of lack of support, lack of commitment, and confusion with regard to their online initiatives. Many local AI groups and members have eagerly volunteered their time and effort to create their own Amnesty Web sites. They want to capitalize on the Web's potential as a tool for outreach and use it to link up with other groups and to promote both AI and their local activities. The Amnesty Lesbian, Gay, Bisexual and Transgender (LGBT) site, www.ai-lgbt.org, is a case in point.

Like hundreds of other AI sites, www.ai-lgbt.org was developed and is maintained by volunteer members of a local group, largely at their own personal expense. However, unlike other AI group sites, the LGBT site was created with the intention of developing a *transnational* space for defenders of the rights of lesbian, gay, bi, and transgendered people to communicate, collaborate, and support one another. Ideally, the site was not to be owned by any one local AI group but was to belong to, serve, and be maintained by an international gay rights community. As a direct result of the site's development, initial contact with international LGBT human rights defenders has been established and a sense of community has ensued. However, keeping these lines of communication open and active has been time-consuming and frustrated by participants' inability or failure to live up to their initial commitments. Site designers had hoped www.ai-lgbt.org would be truly interactive in its support for the international gay rights community. It was expected, for example, that participants and users would contribute relevant articles and documents to the site, where an existing searchable databank would make these easily accessible to the community, thereby generating both discussion and action. However, few, if any, contributions have been made and site designers are despairing—tired of volunteering their own time and effort for naught. Furthermore, the site designers have not been able to convince the IS to actively promote the LGBT site as they had hoped. The IS does provide a directory of links to national section and local group Web sites, but the latter are rarely, if ever, actively promoted.

From the IS's point of view, locally developed AI group sites are largely beneficial to the organization. They provide yet another avenue via which potential members can get involved at a grassroots level. They also allow for members to communicate with each other across national boundaries. The assortment and mushrooming of local AI group sites reflect Amnesty's broad membership base, its loosely federated organizational structure, and its pluralistic, grassroots approach to governance. The result, unfortunately, is a multiplicity of Web sites with inconsistent presences and different ways of presenting and managing information. Although national sections are expected to have developed their own Web site policies and guidelines for their respective members, there is little clarity or consistency at the grassroots level. Members are not always aware of what can and cannot be made available to the public; they may not clearly identify their site's relationship to the official international or national section sites; they may not know when actions have become outdated or inappropriate; site styles and wording may be inconsistent with those of official AI sites; and links promoted by the local site may unintentionally reflect poorly on AI's work.

Therefore, despite the obvious promotional benefits of locally developed AI sites, they do occasionally pose a potential threat to AI's credibility, especially since the organization does not have the resources to commit to "quality control." Legal action may even be taken against the organization as the result of misinterpretations. Consequently, the IS site, www.amnesty.org, encourages local groups to register with the IS, offers links to their sites, and also invites users to comment on these sites so that blatant and harmful misrepresentations can be quickly addressed. This threat of misrepresentation, in addition to the risk of misinformation and manipulation, repeatedly occurs whenever Web-related activism is seriously considered by the organization. These partly account for Amnesty's slow and reluctant embrace of this electronic medium.

No doubt, the Web has the potential to lend itself to public calls to action. It is a transnational, public medium that can be both entertaining and educational. However, there are inherent risks in a Web-based approach to public mobilization: In order to engage with the public, AI must open itself up to it. In so doing, Amnesty risks falling prey to its opponents and to the workings of hackers and "uncivil society."

Amnesty documents and campaign materials have always been vulnerable to abuse. However, the Web has made it easier than ever to copy and doctor documents so as to reflect a contrary stance. Although online petitions may have a particular strategic relevance in a campaign, they can be easily abused in this way. Likewise, it is believed that providing prewritten letters of appeal on an Amnesty Web site may pose similar problems. The wording can be easily changed to reflect an oppositional stance. The letter can then be circulated and used contrary to its intended purpose. The IS and many of the national sections are in the process of reviewing and working to improve these techniques so as to capitalize upon their accessibility without compromising effectiveness.

What cannot be avoided or controlled by Amnesty are independent Web sites that deliberately misrepresent or discredit the organization and/or the human rights movement. However, the same medium may also provide Amnesty with the means to fight back—the means to publicly discredit and undermine attempts at misinformation. A case in point is www.amnesty-tunisia.org. This site, which praised the human rights achievements of the Tunisian government, deliberately misled users by appropriating AI's name. While it did not claim to be an authentic AI Web site, its creators were obviously trying to gain credibility through the adoption of the word "Amnesty" (Whaley 2000, 36). When amnesty-tunisia.org was first brought to Amnesty's attention, AI responded by publicly deconstructing the contents of the offending site: its international site, www.amnesty.org, linked to www.amnesty-tunisia.org in one frame and presented AI's critique in a parallel frame. The inaccuracies and contradictions were highlighted and Amnesty's own position was made quite clear: that the Tunisian government's repression of journalists, political activists, and human rights defenders, including AI's Tunisian Section, remained widespread and that torture and ill treatment in prisons were common.

Yet AI is slow to commit to online activism. This is reflected in the ongoing debate surrounding the online posting of UAs. The current policy states that active UA cases can be posted to AI Web sites, at the discretion of the sections' UA coordinators, on condition that no authorities' addresses are provided and no action recommended.[4] Users who view the case and want to act are instructed to e-mail the UA coordinator responsible for the UA's posting for further details. This enables coordinators to alert the user should a follow-up or correction to the case be issued. Moreover, the user may also be encouraged to join Amnesty and act on behalf of similar cases on a regular basis.

The prevailing thought had been that by not providing the complete details of the UA case the risk of manipulation would be lessened. However, this risk seems rather unsubstantiated when one considers how little control Amnesty has over the cases it distributes to its members. Members consist of those who contribute to AI in some shape or form. Contributions may be financial and/or may include some form of commitment such as letter-writing or participation in a local group. Anyone can become a member of AI. Presumably, therefore, the possibility exists that a fierce opponent of the human rights movement can receive UAs via e-mail and use the information however he desires. A less dramatic scenario is how members, in their enthusiasm, tend to share cases with the general public either by distributing these either in hardcopy or electronically or by posting them to their own personal or group Web sites without AI's knowledge.

Whatever the means of distribution, Amnesty has little control over the public circulation of UAs (and other documents), and e-mail and the Web have greatly increased the extent and speed with which this is done. Amnesty is well aware of the fact that the pro–death penalty lobby in the United States has access to UAs and twists the arguments presented in the UAs in its favor. This largely unimpeded circulation of UAs makes Amnesty more vulnerable to abuse, while increasing its potential outreach. When one considers the power of e-mail, the protection of content on the IS or national sections' Web sites may be more limited and perhaps even more symbolic than real. Amnesty's best line of defense against distortion of information and misrepresentation is a well-resourced offense: vigilance, at all levels of the organization from the local group Web master to the IS. Ultimately, a handful of national UA coordinators are now posting, with the blessing of the IS, a select few full-text UAs online.

Moreover, in addition to being aware of the threat of manipulation and misrepresentation, AI has been reluctant to embrace the Web's great potential for public education and outreach. Amnesty's communication has traditionally been directed toward an internal audience—between members, coordinators, sections, and the IS—and, consequently, e-mail was adopted relatively quickly and has proven to be an invaluable tool. Historically, Amnesty has reached out to the general public almost exclusively via its membership and the media, and with the primary intention of drawing prospective members and donors into the internal stream of communication. Amnesty is now faced with an opportunity (and challenge) to interact directly with the public. The Web has made it possible for AI to dialogue with and invest in the broader public that is more casually concerned with human rights—with people who want to be involved with the organization on a sporadic basis or who simply want to educate themselves without necessarily committing to the cause. With over 250 Web sites worldwide posting information produced by or about AI, and with more than 8,400,000 annual visitor sessions to AI's international Web site alone, not only is the potential for public outreach real, but some human rights advocates have come to expect AI to commit to public education in the virtual world.[5]

However, as important as public education and outreach may be, some have cautioned that focusing on these at the expense of AI content, recruitment, and calls to action may be too costly. In other words, investing in a noncommitting public—a public that takes no AI-sponsored action or makes no financial contribution—may be good for the broader cause but financially disastrous for the organization. Some sections of the movement, especially those that have a history of public outreach, believe that this need not be the case and that ignoring the general public may, in fact, prove to be the greatest cost to both Amnesty and the human rights movement. AI is thought to have some degree of responsibility for shaping the opinion of the noncommitting public—the greatest users of the Internet—and, without a significant AI Web presence, AI is ill equipped to counter online misinformation. Consequently, these sections are beginning to embrace this Internet-based opportunity. They are experimenting with relatively inexpensive and user-friendly interactive applications: audio files including music downloads, video, and short movie clips such as AI-USA's "Conflict

Diamonds" flash movie, which educates viewers about the link between the diamond trade and the brutal conflict in Sierra Leone.

By and large, however, Amnesty continues to lack an educational focus on how local human rights violations relate to broader historical, socioeconomic, and political contexts. The IS still does not view the public as its primary audience. Consequently, the coordination of electronic campaigns and strategies of sections is, at best, confused, particularly as these have widely varying commitments to Web-based activism—commitments that sometimes exceed those of the IS. In the end, most AI Web sites, including the IS's www.amnesty.org, continue to restrict themselves to acting as resource libraries and recruitment agents rather than human rights education portals. In other words, they tend to reflect the pride and bedrock of the organization through investigation, documentation, and strategic mobilization. Users can join or donate to AI online; act immediately on a particular case; gain access to news releases, reports, and documents; learn how to influence legislation in their respective countries; learn how to write an effective appeal letter; or learn of upcoming local events.

Clearly, the Web's public face presents a number of opportunities and challenges to AI. Unlike e-mail, which facilitates internal communication and mobilization between committed individuals, the Web's broader audience drives AI activists and strategists to think beyond the traditional boundaries of the organization, and arguably at the risk of overshadowing forty years of highly respected investigative research and action. Moreover, for a human rights organization that has, historically, attempted to maintain its neutrality in partisan politics and limited its commitment to the realm of civil and political rights, an interactive form of public education presents a formidable challenge.

Additional Barriers to Online Activism: Issues of Representation and Access

ICTs contribute to the deterritorialization of activism by facilitating communication, the dissemination of information, coordination of action, and mobilization of individuals across national boundaries. By the same token, however, ICTs have the potential to further exacerbate and entrench regional and socioeconomic divides. Those who are engaged in online activism tend to represent the elite of the world, which begs the question: Just how grassroots or representative and accessible are these forms of activism?

ICTs do not operate independently of complex and interdependent sociopolitical, economic, or historical contexts, nor does virtual space operate independently of relationships of power. Consequently, ICTs, which may be couched in terms of globalization and democratization or heralded as a panacea, tend to be accessible and beneficial to an elite minority while the majority has neither access nor voice. This so-called digital divide is reflected at both national and international levels. The use and access to the Internet is most widespread and inexpensive in industrialized countries, and within most countries the Internet tends to benefit those who are already socially and economically privileged.

According to the United Nations (UN), industrialized nations account for 15 percent of the world's 6 billion people, 88 percent of whom are Internet users (UN in Bray 2001a). Eighty percent of the world's population, on the other hand, has yet to place a telephone call. Africa has less than 2 percent of the globe's telephone lines, that is, 2.5 lines for every one hundred Africans, whereas there are 70 phone lines for every one hundred Americans (Bray 2001a). More people use the Internet in London than in all of Africa and, there are more users in South Africa than in all other African countries combined (Bray 2001b).

Moreover, those who do have access to the Internet may not have the same means or rights to publish their own views online, as in countries where governments set up firewalls and monitor Internet traffic. Internet users are also presented with volumes of contradictory bits of information, the reliability of which is often difficult to assess. Consequently, despite the vast quantity of information available on the Internet, the quality of political discourse will not likely improve. Also, others will not necessarily welcome the values expressed on the Web—a medium that remains profoundly Americanized. Many parts of the world view the Web as representing consumerism, lax morality, and the unrelenting drive of American culture. Ultimately, the assumption that the Internet or ICTs are inherently inclusive and representative by virtue of their transnational nature continues to be one that is clearly unfounded.

As a means to begin to bridge the digital divide, Amnesty has committed itself to a multilingual approach to ICTs. Since one of the major hurdles to access is language, one of Amnesty's short-term Internet strategies is to provide links to international AI sites in Arabic, French, and Spanish as well as to national or section sites that operate in less widely known languages. There is also a push toward establishing and maintaining regionally specific or thematic discussion groups in a number of languages. The more linguistically accessible AI is to the general public, the more inclusive it is likely to be as an international organization.

Amnesty is also committed to improving of its information technology support to sections. Most, if not all, of AI's poorest sections face exorbitant costs for hardware, Internet access, and telephone services. Combined with weak infrastructures and a need for skilled personnel, many sections are struggling to get online. Priority is now being given to ensuring reliable access to equipment and technical support. Resources garnered by the IS and wealthier sections are pooled and redistributed primarily according to identified needs, but also according to the strategic importance of a particular section. In other words, there may be heavier investments in sections that have the potential to wield a considerable amount of regional influence. In addition to this formal approach to resource redistribution, there is also some sharing of technological resources occurring informally, between AI sections. For example, a local Canadian Amnesty group donated a personal computer (PC) to AI's Jamaica section—its first and only PC to date.

However, the acquisition of equipment is no guarantee that a section will have the capacity to engage actively in transnational online activism as unreliable telecommunication infrastructures and the exorbitant cost of ISP services continue to limit access to the Internet. Although AI is working to standardize technology across the movement to facilitate communication and increase speed of action, it recognizes that this may not be immediately feasible for a number of sections. Until (if ever) these limitations are addressed, AI will continue to use a variety of communication tools so as not to risk excluding a potential or existing section, for if ICTs inadvertently contribute to an overrepresentation of some countries over others, Amnesty may be (justly) discredited as a globally representative movement and its effectiveness may ultimately be compromised.

Organizational Challenges Raised by ICTs

AI has become increasingly reliant on information and communication technologies to carry out its activities internationally. As beneficial as some of these technologies have proven to be, others have been approached with trepidation and all have challenged the organization to rethink the shape and future of activism. However, the impact of ICTs has not been limited to strategy and action. Rather, these have had significant implications for AI's resources and for its managerial or organizational culture.

Wiring, upgrading, and maintaining AI's technology is an expensive endeavor. The organization's commitment to standardizing technology across sections and supporting them requires a significant financial investment. There also needs to be an investment in training of existing staff and/or volunteers—time and effort that are often considered to be taxing on already burdensome workloads. On the other hand, the heavy investment in technology is also expected to "pay off." The cost of introducing new technologies is likely to be offset by the ability to recruit electronically and support more donors and members. Moreover, many programs, including the UAN, are beginning to see a decrease in discretionary costs of operation (such as postage, stationary, photocopying, or faxing).

The permeation of information and communication technologies at all levels of AI has also had a profound and lasting impact on Amnesty's managerial and organizational culture. Some age-old problems are exacerbated and new ones have emerged. No doubt, the means of communication have vastly improved, largely because of the widespread adoption of e-mail. However, actual communication and coordination between sections and structures remains, as always, awkward or disjointed. This could be accounted for by AI's fragmented organizational nature. Structures have evolved to respond and reach out to particular constituencies—some localized, others transnational—but continue to have a poor record of communicating with each other. Consequently, the development of internationally representative or joint strategies also continues to be difficult. The Web, in particular, is challenging AI to better integrate these relationships so that its public face will be one that is consistent and clearly defined.

This difficulty is compounded by the fact that Amnesty is gradually being drawn into becoming more reactive than it has been comfortable doing in the past. Historically, AI has established its positions and action agenda based on a lengthy research and approvals process. Now, the media and the public have come to expect that it will offer immediate commentary and opportunities for action on issues that, through the media, have captivated the public's imagination. In other words, a clear, common front must be agreed to quickly and made public with equal speed.

On the other hand, the membership is pulling Amnesty to respond to *its* evolving needs—needs that are also influenced by the opportunities new technologies afford it. For example, AI members who are electronically drawn together by a particular theme and/or language require some form of support, but it is not clear what kind of support or who, within the organization, is to provide it. Once again, Amnesty's loose organizational structure, or "organized anarchy," and the possibilities afforded to it by the Internet place competing demands on that very same structure. Consequently, the IS and AI sections are having to reinvent themselves. Managerial responsibilities, program areas including membership support and campaigning, and fundraising are all having to be redefined in direct response to members' changing needs. Again, this is no small feat for a large, unwieldy, and disaggregated organization that, traditionally, has been resistant to profound change.

In addition to having to respond to a plethora of needs and competing priorities there are also important decision-making procedures that have yet to be established. Who makes decisions pertaining to the electronic medium? Is it the managers, who may not be familiar with the medium? Or is it the information technology staff, who may be more at ease with the mechanics of technology than with their potential strategic use? Then again, is it the campaigners or program staff, whose job it is to strategize and orchestrate action? Is AI to cultivate staff and volunteers with new sets of skills, or has it come to expect that these individuals will have some basic experience using technology? Clearly, there are no straightforward answers to these questions as the experience and resources of each section varies so widely. However, as ICTs become increasingly integrated into Amnesty's broader mission, all levels of AI must, at some point or other, seriously engage in these debates.

Finally, the advent and widespread use of information and communication technologies is affecting Amnesty in yet another unexpected way. Some of those that AI purports to represent now have the means to speak for themselves. Human-rights defenders, social justice advocates, and local NGOs are actively using the Internet, the Web, and e-mail, in particular, to access the general public directly and often do so more effectively than Amnesty. This development, whereby smaller human rights NGOs are appropriating a louder voice, is applauded by AI and is truly a coup for the human rights movement. Yet the organization is left to ponder what this may mean for its role in the movement and for its relationship to local human rights defenders. Moreover, how will the public come to view and support AI's work in light of this development?

There is a negative aspect to this levelling effect of the Internet, as well—one that is of great concern to Amnesty. All online activists, regardless of their cause, standards, legitimacy, and systems of accountability, are vying for equal space on the Web. Consequently, many tend to be viewed equally by the media and by the general public. Organizations like AI and Human Rights Watch, which have prided themselves on years of careful documentation and research, are now lumped together with groups whose mandates are entirely conflicting with AI's. In other words, civil society's heterogeneity may be leveled out by, and inaccurately represented in, cyberspace. Amnesty's challenge is to distinguish itself from groups that could potentially tarnish its commitment to human rights and, ideally, to reach out to those with whom it could collaborate as a means to further its cause in a meaningful and truly international or representational form.

Negotiating the Challenges of ICTs

Communication is absolutely fundamental to an organization, such as Amnesty, that is membership based. The advent of information and communication technologies has vastly improved the speed with which data can be collected by AI and circulated internally and publicly. In turn, this has facilitated the coordination of action and the mobilization and delivery of organizational support to activists. However, ICTs also aggressively challenge AI to interact directly with the public in a still unfamiliar, unbounded, deterritorialized, and transnational space. In response to the changing needs and expectations of its membership and the general public, Amnesty's online presence must extend beyond the comfort of a simple resource library or recruitment agent to become the leading online source for human rights. This is particularly important when one considers the leveling effect of the Internet, with various groups (or factions) within civil society, including opponents to the human rights movement, competing equally for virtual space. If AI is to adopt this new online persona successfully and if it is to be flexible and open to the changing nature of human rights discourse—nurturing a multivocal shaping of the international human rights agenda—it needs to make itself easily accessible and take concrete steps to bridge the digital divide. In so doing, however, Amnesty may be rendered more vulnerable to misinformation, misinterpretation, and manipulation. Paradoxically, therefore, in order to open itself up to the public, Amnesty must be more protectionist—tightening its security controls and becoming ever more vigilant to occurrences of abuse. Having experienced both the benefits and the limitations of information and communication technologies, Amnesty has come to favor a utilitarian or pragmatic approach, rather than one that is particularly visionary. ICTs are a new communication medium and not a new world. They are most useful to AI when used in conjunction with other more traditional forms of communication, most importantly including face-to-face interaction.

Moreover, while the introduction of new technologies may have improved the medium they may also have posed unexpected organizational and financial challenges to and strains on AI. In particular, the existing structural gaps in Amnesty's fragmented organizational framework appear to be simultaneously bridged and widened by technology: ICTs facilitate communication between AI structures all the while nurturing expectations that Amnesty should offer both immediacy of action and even greater flexibility. In other words, online "communities" or networks of members are demanding more flexible support.

And while AI's means of communication may have improved, there is now a need to communicate even faster as it is expected to respond to crisis with even greater immediacy.

Ultimately, information and communication technologies have become an integral and valuable element of AI's strategy and commitment to the respect of international human rights. However, given the complexities embedded in the international human rights movement and the equally complex demands of those who contribute to it, technology cannot be accepted at face value but, rather, must be used strategically, as are all other campaign elements. Their limitations must be acknowledged and the challenges they pose to the organization must be identified if ICTs are to be effective, locally appropriate, and inclusive across territorial and institutional boundaries. As long as these limitations and challenges are borne in mind, information and communication technologies are likely to continue to benefit Amnesty and its advocates.

Notes

1. Amnesty International works to promote the principles enshrined in the Universal Declaration of Human Rights. It has more than 1 million members and supporters representing 162 countries and territories. AI acts impartially and independently of any government, political persuasion, or religious creed, and it is largely financed by subscriptions and donations from its worldwide membership. As a grassroots organization, AI is accountable and responsive to its members, who have voting rights within the organization; they have been the main drivers of its direction and focus over the years.

2. Amnesty's organizational structure is notoriously fragmented. Generally speaking, "sections" refer to a national representative body of Amnesty. A section may be managed by paid staff members with the support of volunteers, or, if resources are limited, it may be run exclusively by volunteers. Sections tend to act as intermediaries between the IS and local groups, networks, and members.

3. In previous years, up to one-third of UA cases have had positive developments.

4. AI-USA and AI-Sweden are, to date, the only sections to openly post UAs online in their full text forms, including both recommended actions and authorities' addresses.

5. www.amnesty.org is expected to host over 8,400,000 visitor sessions in

2001—a projection based on the first five months of 2001—up from 1,400,000 in 1998. Although the numbers are truly impressive, the increase is entirely in line with the estimated general growth of internet use. In other words, Amnesty's popularity online has been maintained. It has neither increased or decreased in proportion to the Web's overall traffic (CCWG 2001).

References

Bray, H. 2001a. "The Wiring of a Continent: Africa Goes Online." Online. http:///dailyglobe2/203/nation/Africa_goes_online+.shtml .

——. 2001b. "The Wiring of a Continent: Entering the Queue." Online. http:///dailyglobe2/204/nation/Entering_the_queue+.shtml.

Computer Communications Working Group (CCWG). 2001, June. *Meeting Notes*. London: Amnesty International.

Whaley, P. 2000. "Human Rights NGOs: Our Love-Hate Relationship with the Internet," pp. 30-40 in *Human Rights and the Internet*, ed. Steven Hick, Edward F. Halpin, and Eric Hoskins. London: Macmillan Press.

10

Ethnic Online Communities

Between Profit and Purpose

Steven McLaine

"Community is quite possibly the most overused word in the Net industry."
—Janelle Brown (1999)

By now, the arguments on both sides of the debate over the validity of online communities have been well established. Some researchers believe that online communities are legitimate forms of community, either as extensions of real-life communities or a new breed of community altogether (Rheingold 1993; Wellman and Gulia 1999); others are skeptical of their impact on individuals and society as a whole and question whether they actually deserve to be called communities at all (Lockard 1996; McClellan 1994). I find particularly interesting a tangential perspective—the inherent potential for commodification of online communities. As I discuss in this chapter, the consequences of imbalanced choices between profit and purpose are unfortunately and especially magnified in many ethnic-specific online communities, hampering any meaningful and actual online or offline community empowerment for the users that could benefit most.

As soon as online communities were first created, their economic viability was debated (Armstrong and Hagel 1997). Years later, these communities have become "big business," spawning new products such as *The Online Community Report. The Report* is an e-mail newsletter created to help online community creators deal with the trials and tribulations of maintaining their "investments," including shrinking venture capital and fickle community residents. Various articles share online community best practices; one editorial in particular by Senior coeditor Jim Cashel (2000) calls for brave souls to provide "economic footing" for the sector by initiating subscriber fees. Yet another by Cashel (1999) is simply entitled "How to Sell Your Online Community," which reveals that certain online communities have already been sold for millions of dollars.

Profit and community make curious bedfellows. As Janelle Brown (1999) noted in the online magazine Salon.com the temptation to sacrifice necessary community maintenance and development in order to ensure financial return seems to have proven difficult to resist, judging by the glut of impotent community providers such as GeoCities, Xoom, and theglobe.com. Rarely, if at all, do these "community centers" inspire or facilitate efforts toward individual or group empowerment, organizing, or advocacy. Simply typing the word "COMMUNITY" in big HTML letters doesn't make it so. Even Rheingold (2002), a pioneer of studying online communities, warns on his Web site (www.rheingold.com/vc/book/intro.html) " . . . you have to be careful to not mistake the tool for the task and think that just writing words on a screen is the same thing as real community."

To this troubling mix I will now add race and ethnicity, simply because few others have in past cyberculture research. Burkhalter (1999) explained how racial identity is expressed in Usenet discussions, but popular new ethnic online communities (EOCs) remain a fertile ground for debate and analysis. Ethnic Internet portals such as BET.com, Yupi.com, and A Space.com have received major attention by marketing the premise of relevant content and community to underserved Internet users of specific ethnicities. The EOC market is booming precisely because traditional Web programming has typically ignored minorities. Silver (2000) notes that the Blacksburg Electronic

Village—a community network in Blacksburg, Virginia—"routes around" race and relies instead upon the "digital default": white, male, heterosexual, middle-aged, middle-to-upper class (143).

Referring to this malaise of disenfranchised Netizens, Nakamura (1999) dismisses the utopian notion of the Internet as a "social leveler," considering that its neutrality is only valid because issues of race and ethnicity are avoided, rather than accepted. Both she and Kolko (2000) refer to the Internet's default status of "whiteness." To further this notion, Nakamura also finds that many users that dare to acknowledge racial or ethnic characteristics in the LambdaMOO chatspace are often accused of engaging in "hostile performance." Certainly this concept of performance can be extended to the Internet as a whole. The anonymity of the Internet can work both ways. True, no one can see what color I am, but *no one has to see what color I am*. Therefore, the touchy subject of race can be brushed under the mousepad.

Ignoring issues of ethnicity limits the growth of individual users, as well as the technological medium as a whole. Most certainly, this lack of recognition—or even acknowledgment—of ethnic Web users has facilitated the infamous digital divide. Although other factors such as income and education also contribute, there is still a significant disadvantage in access to technology for African Americans and Hispanics attributed directly to race. The National Telecommunications and Information Administration report *Falling through the Net: Towards Digital Inclusion,* released in October 2000, states:

> Differences in income and education do not fully account for this (racial) facet of the digital divide. Estimates of what Internet access rates for Blacks and Hispanic households would have been if they had incomes and education levels as high as the nation as a whole show that these two factors account for about one-half of the differences. (U.S. Department of Commerce 2000)

Certainly, the "digital default" status serves as a deterrent to even those minorities with access to and interest in technology because they perceive a lack of reward within the Web because of its limited ethnicity-relevant community and content. In March 2000, The Children's

Partnership (TCP) published a report entitled *Online Content for Low-Income and Underserved Americans: the Digital Divide's New Frontier*. TCP performed an audit of current online content and compared it to the needs and requests of divide "have-nots" who wished to become acclimated with technology's vast resources. TCP found that individuals that have low income, live in rural communities, have limited education, or are members of racial or ethnic minorities are interested in four types of content:

1. Sites for limited literacy or multilingual readers
2. Local job resources or job listings for entry level positions
3. Local low-cost housing information
4. Local cultural information (The Children's Partnership 2000)

Their conclusions seem to deviate severely from the direction of these universal ethnic community sites. The report mentioned the Web's trend to instead focus on general cultural information, and listed these concerns with ethnic online communities:

> However, while most build important bonds that can tie members of ethnic groups together, it is often difficult to do so without ignoring some of what makes certain ethnic cultures unique. The trend toward homogenizing and ignoring differences, dictated by a desire to build market share, shuts out to some extent the distinctive essences that give vitality to a culture. For instance, A-Space.com is working to bring together the Asian world under one umbrella, a difficult challenge because of the widely varying heritages among Asian Americans. Yet it is precisely these distinctive differences, which are often lost in big portals, that can provide great impetus for ethnic groups to actually go to the online world. These unique traits are also what can bring people into a community access center to develop culturally relevant information together. (The Children's Partnership 2000)

The lack of online outlets for Web residents who are low income and/or people of color presents an incredible opportunity for potential EOC creators. Obviously, there is potential for financial gain by servicing minority users. In the past year, for example, African Americans and Latinos have become the two fastest growing markets for Internet access in the United States (U.S. Department of Commerce 2000). (These statistics are based on the fact that Blacks and Latinos began the year with the lowest number of homes with Internet access; however, the numbers are still revealing.)

Beyond financial gain through serving minority users, legitimate EOCs could help empower those marginalized individuals traditionally ignored by technological advances. Any online community should feel responsible for honoring the voices of its users, but ethnic communities even more so, because the stakes are that much higher. Helping communities adapt for the twenty-first century is indeed significant, but providing a platform for groups typically unheard and unrecognized in the Internet medium could prove quite daunting.

The traps and pitfalls that would accompany such a venture are many. As Sterne (2000, 194) wrote, we should not "mystify the medium." The Internet, as it now stands, should never be mistaken as an "equalizer" while access to technology remains limited for a significant number of individuals of all races and ethnicities. Simply advertising an EOC would not effect real change among underserved users. Offering incentives, however, for "forgotten" ethnicities to investigate technology could inspire increased interest.

In addition, an ethnic online community assumes success in a model long ago abandoned in real life. A divide existed before the word "digital" was attached to it. The emphasis on technology has only served to reveal the many inequalities that exist in this country. No community that exists today is based purely on race or ethnicity. Class is a major factor. There are gated communities of color now, as well, and those gates are equal opportunity repellants.

With the aforementioned factors in mind, I examined the process of providing three ethnic-based online communities for Asian Americans, African Americans, and Hispanic Americans to determine the influences of both profit and purpose. Community Connect Inc. (www.communityconnect.com) operates three of the most popular online communities for "U.S. ethnic audiences"—AsianAvenue.com, BlackPlanet.com, and MiGente.com. By following their online offerings, I hoped to see if this cookie-cutter approach could produce effective communities (contrary to the concerns of The Children's Partnership), and if/how they sought to balance their goals between profit and purpose.

I wanted to know if, in fact, they would acknowledge their responsibility to the communities they claim. (As this was my first actual membership experience in an online community, I felt confident I would harbor no unfair expectations based on other communities offered by Internet competitors.) And through this comparison, I hoped to generate a blueprint of essential characteristics that any potential EOC should follow in order to have relevance and impact.

Locating Ethnicity Online

Launched on July 21, 1997, AsianAvenue.com was the company's first online community/venture. After receiving acclaim and awards (as well as a large number of members), AsianAvenue.com's success subsequently encouraged Community Connect Inc. to continue expanding into other ethnic markets. BlackPlanet.com followed in September 1999, and MiGente.com debuted in October 2000. BlackPlanet.com and AsianAvenue.com boast over 5 million and 2 million members, respectively, on each of their Web sites; with more than six-hundred thousand members, MiGente.com hopes to follow in their footsteps and become the premier Latino site in the United States (www.community connect.com 2001).

Prior to their triumphant debuts, however, Community Connect Inc. CEO Benjamin Sun initially found it difficult to convince venture capitalists that a market even existed for AsianAvenue.com and other ethnic online communities (Yang 1999). These early funding struggles illuminate further the prevalent dismissal of race and ethnicity when considering an audience for potential Internet initiatives. Still, Sun was confident that the need existed for race-based online communities. He explains, "Ultimately, the online world is based on people interacting and establishing relationships. There must be some shared interest and background. What's more powerful than race and ethnicity?" ("Wired Like Me" 1999). In another interview, Sun adds, "Communities, as we define them, have the ability to allow people to build relationships around the most compelling tie: race and ethnicity" ("Casting Wider Net" 1999). And in yet another interview, he notes, "A Web site for car lovers is not a community. Age, religion, ethnicity; those are the ties that bind together a community" (Ramirez 1998).

In explaining why he decided to create AsianAvenue.com, Sun revealed what he believes to be key components of an online ethnic community:

> I saw what people considered online community but they weren't really meeting my expectations. It was static, Webzine-like content with message boards and chat rooms thrown in. But people need to be empowered and people need to drive the community. ("Virtual Enclave" 1998)

His thoughts are echoed in Community Connect Inc.'s mission, as detailed on their Web site:

> Our proprietary technology gives our members the tools that enable them to form and engage in culturally-oriented communities. More than our content, however, our sites offer our members instant communication and networking opportunities, resulting in a highly personalized and meaningful community experience. . . . Our members repeatedly tell us that "their" site has become an integral part of their lives, online and offline. (www.communityconnect.com/aboutus.htm 2001)

So these three EOCs were initially created to reflect and endorse the tenets of "relationships," "activities," and "empowerment." How, though, do you evaluate the success of a community with such lofty goals? On BlackPlanet.com's login page, a quote from a satisfied member reads, "I see this site setting the foundation for a strong online Black community." Despite this ringing endorsement, creating benchmarks for measuring the social impact of communities can be difficult. It is much easier, in fact, to determine success by more conventional means.

In the October 1999 edition of the *Online Community Report*, senior coeditor Dan Shafer lists the "most relevant metrics" for measuring online communities—unique visitors, page views, registered members, postings per month, time on site, posting ratio, and audience participation. Indeed, BlackPlanet.com proudly proclaims on its login page that the *Wall Street Journal* reported Black Planet as the "stickiest" Web services site. The average amount of time per user spent at Black Planet.com was an impressive 34.4 minutes ("Reality Bytes" 2000). And in fact, another quote posted to the login page upon a subsequent login exclaimed, "I never thought I would spend so much time at a site as I have at BlackPlanet!"

In addition, Community Connect Inc. reports "very substantial ad revenue" for both AsianAvenue.com and BlackPlanet.com, with MiGente.com soon to follow if projections are correct. Here the dynamic of profit versus purpose is made evident. Profit is easier to measure and benefit from, while purpose is more elusive and indirectly rewarding. A company cannot pay staff with empowerment. Ultimately, these types of decisions can easily contribute to a realignment of the EOC creator's priorities—to the detriment of the communities they sponsor.

A proven formula for success was established with AsianAvenue.com, and the other two Web sites do not deviate far from the norm. Each site offers personalized Web pages, chat rooms, forums, e-mail, games, and prizes for frequent visitors. Community Connect Inc. focuses on providing the structure for the "community" and relies heavily on user input for site content. Community members usually provide the topics highlighted for forum discussions and are also encouraged to write in and respond to any original content offered by the respective sites. Community Connect Inc. Vice President Calvin Wong explains the strategy:

> It's the people, the vehicle for communication you present them with. You offer them high-level avenues of interaction. Give them autonomy and membership. You say to them, "You get to set up the content. You run the community." (Gach 1998)

Membership is free, requiring only the completion of a detailed questionnaire to create a user profile. Essential information such as name, location, gender, date of birth, relationship status, household income, highest education level completed, college attended, year of graduation, industry, and occupation are universal throughout the Community Connect family of sites. Each site also asks for Interests and Background of the registering user, including Sexual Orientation, Music Tastes, Sports, News, Arts, and Recreation/Hobbies. Users can determine which aspects of their profile to feature on their personal Web page (which I thought was a nice touch).

Here, slight differences among the Web sites begin to emerge in accommodation of ethnicity. For example, MiGente.com is the only site of the three that inquires about religion; it also requests classification of both race and ethnicity. Members can choose from twenty-three countries for their ethnic origin, ranging from Argentina to Spain. Options for musical tastes include merengue, salsa, tejano, and world music. When I joined MiGente.com, I was also asked if I preferred speaking and reading in English, Spanish, or both.

BlackPlanet.com asked only for race, while AsianAvenue.com asked only for ethnicity. For ethnicity, the latter site offers seventeen options, including Chinese, Indian, Thai, Filipino, Pakistani, Vietnamese, and Korean. Upon the site's inception, creating these ethnic options hinted at the inherent difficulties of supporting an EOC. After compiling an initial list of choice for ethnicity and going online, the fledgling site was immediately flamed by a group of Asian Americans located in Minnesota. They wondered why the Hmong had been omitted from the list ("Wired Like Me" 1999). CEO Sun and his executives couldn't really complain; much of AsianAvenue.com's success can be attributed to the fact that second-generation Asian Americans are usually scattered across various geographic locations, making it difficult to organize real-life community (Lee 1999).

However, the concerns raised by TCP speak directly to this dilemma. The different ethnicities all grouped under the Asian heading each possess unique values and cultural beliefs. Lumping them all together only dilutes the differences that might actually inspire new ethnic users. In fact, some of these ethnicities have their own EOCs, for example, Indolink, the top-rated online portal for the Indian community.

AsianAvenue also offers a Lifestyle option, with choices of Kid, Teen, College, Graduate Student, Young Professional, Parent, Senior, and Gay/Bisexual. Additional musical taste options included Asian pop. In addition to providing fellow site residents with information, these individual profiles also enable, in Community Connect Inc.'s own words, "tribal" marketing—as featured in AtNewYork.com's "10 Business Plans to Watch in 1999" (1998). Readers will have to excuse my naiveté, but I was both amazed and disappointed to visit the Community Connect Inc. Web site and see statistics on all of the members of AsianAvenue.com—broken out by the information I had just provided in registering weeks ago. Neatly compiled and presented for potential clients were percentages of AsianAvenue.com residents by gender, age, lifestyle, income, and state of residence (www.community-connect.com/mediakit). I assume that also available, if companies are interested, are lists of residents' favorite sports and hobbies, preferred musical tastes, and other various interests.

In fact, the registration processes for AsianAvenue.com, Black Planet.com, and MiGente.com have produced a veritable gold mine of niche marketing information. Community Connect Inc. estimates that Asians, African Americans, and Latinos together make up nearly one-third of the U.S. population and possess over $1 trillion in annual purchasing power (www.communityconnect.com/mediakit). In addition, Web users generally spend four to six times longer at online communities than other Web sites, ensuring adequate attention for strategically placed ads ("Economic Viability of Web Forums" 1997)—hence the excitement over the *Wall Street Journal*'s "sticky" award. Anya Sacharow, an analyst from technology research firm Jupiter Communications, agrees, "We basically have argued that by targeting an affinity group, a community becomes more valuable for members and consumers online" ("Casting Wider Net" 1999). The communities of these Web sites are, in effect, targeted and captive markets for advertisers.

And companies take full advantage. Members log in to find that the amount of ads is easily equivalent to site content, if not surpassing it. Banners for Chase and MTV run simultaneously through all three community sites. While the text for the MTV ad is the same, the accompanying faces change to reflect each respective EOC. Other advertisers include Spiegel, Snap-On Tools, Hookt.com, Physique, and Citibank. I am fairly certain that these ads do not empower anyone in the EOCs besides Community Connect Inc. Of course, most Web services support themselves with advertising, particularly when offering free services. However, a company that claims to represent ethnic communities could be more selective when choosing advertisers. None of the companies featured are notably minority-friendly. For example, most of the ads at BlackPlanet.com showcase online banking and credit cards. Meanwhile, banking loan practices have been repeatedly criticized as unfair within the African American community (see Hudson 1996). Credit cards and the personal debt they help to encourage are not particularly empowering, either.

Clearly, profit has taken priority over purpose. During the fourth day of Kwanzaa (an African American spiritual holiday that celebrates seven principles that emphasize unity within the family and community), a miniscule banner with squint-necessary text was posted on the main page of BlackPlanet.com to "honor" Ujamaa, or Cooperative Economics. Since BlackPlanet.com did not bother to post the actual meaning of that particular principle, I will do so here. Ujamaa means "to build and maintain our own stores, shops and other businesses and to profit together from them." Ironically and unbelievably, the five-word banner was dwarfed by an absolutely immense ad directly above it offering cash back from Ford Credit on a new Harley Davidson. Also posted was a picture of the Kinara, which is a candleholder that represents the seven principles of Kwanzaa, and a picture of money. I am still not sure if that was for Ujamaa or Community Connect Inc.

I did notice one offer to "give back" to the community—Free Lotto.com advertised on all three sites that it would donate fifty cents to charity each time a Community Connect Inc. member signed up for FreeLotto. On BlackPlanet.com, they offered to donate to Project SHE, an organization that provides breast cancer education to African Americans worldwide. On AsianAvenue.com, the selected charity was Asians for Miracle Marrow Matches, and on MiGente.com, they offered donations to a nondescript "charity that helps the Latino community." Despite this uneven attempt to find relevant charities (although breast cancer and bone marrow disease are indeed worthy causes), as well as the fact that gambling has contributed to dearth and decline in all minority communities, the blueprint remains valid. EOCs can and must be financially responsible to the communities they represent. At the very least, EOC creators could use some of the ad money to provide outreach to those members of minority communities that lack access to technology.

While attempting to serve the two masters of community and commodity may bode differently for each ethnicity, the struggle to be heard remains common. BlackPlanet.com's CEO Omar Wasow laments, "My community, the black community, has a deep urge for a voice . . . especially since we haven't had a voice in traditional media" (Lee 1999). Wasow believes that the existence of BlackPlanet.com and other ethnic online communities such as NetNoir.com, BlackVoices.com, and BET.com deflates concerns about a "racial ravine" in Internet access:

> Part of what's exciting about [Black Entertainment Television's partnership with Microsoft, News Corporation, Liberty Digital Media and USA Networks to create BET.com] is that it validates the market. John Malone [CEO of Liberty Digital Media] and Rupert Murdoch [CEO of News Corp.] believe there are black people online. So there's no reason to buy into the digital-divide cliché that there aren't any black people online. (Lee 1999)

Here the intermingling of community and business becomes dangerous. As the executive of an online community, it is essential to convince potential investors of the strength of the existing marketplace, that is, online African Americans. However, as the facilitator of a community—and especially as someone who professes to represent the voice of the African American community—it is quite irresponsible to label the digital divide as a cliché. Despite its optimistic title (no doubt chosen to reflect the new administration's digital divide approach), the latest NTIA report—*A Nation Online: How Americans Are Expanding Their Use of the Internet*—indicates that 60 percent of all African Americans and 68 percent of all Hispanics still lack access to the Internet (U.S. Department of Commerce 2002). The divide does exist, and any community—online or offline—that seeks to empower or provide a "voice" must first acknowledge its underserved members, rather than relying on Mr. Murdoch's assessment of the "market."

As I enter the respective EOCs, I am wished "Merry Christmas" ("Feliz Navidad" on MiGente) and am immediately greeted by the usernames of other community members. At AsianAvenue.com, I see Fine AssAznGurl, QtLilaZn_pNyGr, ThugRidah_80, Fili_Dragon, and hoocheemama, among others. Online at MiGente.com are mami928, blacklatindiva, spicygirl, AlatinoLover, and sexeyrikan. Dollabillyall, blackbold1, soulsista, gambianqueen, and brown_koffee are just several of the members waiting to chat at BlackPlanet.com.

I immediately think of Nakamura's (1999) discussion of "identity tourism," and how she decried the use of such stereotypical appellations by Caucasian users as a perpetuation of negative discourse. Yet here we have individuals within their own communities freely flaunting stereotypes as identification. In essence, some of the tour guides may be taking the tour themselves for fun. Is this residue from existing in a medium where acknowledgment is rare? Or perhaps the vehicle of the Internet itself still elicits an "other" reaction from traditionally unrecognized participants, even though the EOC was designed for such recognition?

There are theories amidst the African American community about the use of the word "nigger" and how its use by African Americans only spites and defies its negative connotation. Many within the community have adopted the word, but there and only there is its use condoned. By speaking the word and determining its existence, there is a sense of controlling the meaning as opposed to its intent of insulting African Americans. In this case, perhaps, we see a glimpse of the empowerment mentioned by CEO Sun as a key component in creating ethnic communities online. If so, can the EOC take credit? Allowing the opportunity to create user names does not necessarily produce responsibility for the thought process behind the name selection. And in cases when it does, is there a means to harness this energy in a manner that benefits all of the community? In a way as effective as the tables of personal information marketed out to companies? These vehicles must exist if the community is to be complete and relevant, but it appears Community Connect Inc. may be focusing on different priorities.

While each site follows a similar format, subtle differences separate the Web sites and communities. The color scheme for BlackPlanet.com is a subdued gray and blue, while AsianAvenue.com greets users with bright shades of green and yellow. MiGente.com, meanwhile, offers vibrant oranges and browns. The main page for all three sites is the same, with sections for News, Channels, Forums, Community, a spotlighted Member of the Week, an occasional soap opera, and a Poll. The site content is a perfect example of the contrast between profit and purpose. Polls range from thought inspiring ("What did the Supreme Court election decision mean for black voters?", "Do you think Latinos are more vocal than other people?") to functional ("How often do you use your AsianAvenue.com e-mail?", "What factors most influence your decision to contact someone at MiGente.com?", "Do you own a computer at home?").

However, the content provided on each site certainly caters to its intended audience. While major headlines are uniform throughout, ethnic-specific information dominates the sites. I frequented these Web sites during the resolution of the Bush/Gore election drama, so each main page included a news item regarding Bush's eventual victory. The similarities quickly ended there. Other pressing news items for AsianAvenue.com included "South Korea's Kim Accepts Nobel Prize," "Wen Ho Lee Case Debriefings Begin," "Korea Talks in Stalemate," "Increase in Asian Males Grocery Shopping," and "Award to Celebrate 'Japanese Schindler.'" Headlines at MiGente.com read "Putin, Castro Slam US Embargo," "Hispanics Less Likely to Get Surgery," "Nader Not Apologizing," and "[Jennifer] Lopez a Spoiled Princess." Black Planet.com offered "3 Arrested for Hate Crimes in Fla," "AOL Time Inc. Is Born," "Popularity of Black Santas Grows," "Powell Named Secretary of State," and "Blacks Show Little Faith in Bush."

The most popular category of forums (as denoted by number of user threads) for BlackPlanet.com, MiGente.com, and AsianAvenue.com respectively are Sex and Love, Relationships, and Money—by fairly wide margins in each case. The similarity between popularity in forums between BlackPlanet.com and MiGente.com can be explored through prevalent discussions on HIV/AIDS that took place concurrently on each site. On MiGente.com, the discussion was entitled, "Why is the Latino community so silent on this issue?" One member responded:

> Why are We Silent??? In addition to all the great opinions, I believe that one of our issue is that la Gente no quiere HABLAR sobre el sexo. How many of you had your parents sit you down to discuss Sex and how to protect yourself or to show you how to use a condom, etc.? It begins at the roots, if it wasn`t discussed openly in the house, how comfortable will you be talking about it to your partner? If your parents are not comfortable talking about the sex, they probably are not comfortable in discussing sexually transmitted diseases incl. HIV. Their parents probably did not talk about such issues, and so it becomes a never ending cycle. I am an HIV/AIDS educator, in the year that I have been doing this for my community, I`ve seen a lot complacency on this issue.

A similar discussion on BlackPlanet.com took place under the heading, "Why is the rate of infection so high in the African American community? What can be done to prevent its spread and help?" One posted reply read:

> THANK GOD FOR BLACK PLANET INFO. THANK GOD FOR THE INFORMATION AND COURAGEOUS INDIVIDUALS THAT BRING THEIR STORIES TO LIGHT ON THE WEB SO THAT OTHERS MAY LEARN AND TAKE PRECAUTIONS. WE NEED MORE FORUMS AND AVENUES FOR DISTRIBUTING THIS IMPORTANT INFORMATION. PEOPLE SOMETIMES NEED GUIDANCE AND PREVENTION AGAINST THIS DEADLY SCOURGE. UNTIL WE, AS A COLLECTIVE WHOLE TAKE RESPONSIBILITY FOR OUR ACTIONS, WE WILL ALWAYS BE HEARING OF TRAGEDIES SUCH AS THE ONES WE ARE READING ABOUT. PLEASE, LOVE YOURSELF FIRST, BEFORE TRYING TO LOVE SOMEONE ELSE. {CONCERNED}

In addition to like-minded sentiments, these responses indicate both the strong need for effective EOCs and the fervent expectations of their members. Each refers to their "community," and they are obviously speaking in terms greater than a Web site. The BlackPlanet.com member pleads for "collective responsibility" and "more forums and avenues for distributing important information." The information has already been distributed to the BlackPlanet.com community; clearly, there exists a greater agenda. EOCs should have mechanisms in place to expand the discussions and resources of its members to the greater offline community. The MiGente member complains that "people don't want to talk about sex," but they do online—buttressed no doubt by anonymity. It's a start, at least, and the issue is so important that any and all potential avenues (no pun intended) to solution must be exhausted.

The potential for offline empowerment inspired by online activity does exist; I will acknowledge that many exciting and worthwhile community ventures have been inspired through these EOCs. The residents of AsianAvenue.com, for example, have effectively organized around various issues relevant to their ethnicity on several occasions, offering ammunition for executives that claim online community can be just as effective as real life. The Moy family came online to seek a bone marrow donor for their daughter Cindy because she was dying of leukemia. The AsianAvenue residents galvanized around the crisis, creating a special forum for her and organizing visits to blood donation centers. A matching donor was eventually found in Singapore; unfortunately, it was too late. However, the story lives on as an example of the power of online community ("Wired Like Me" 1999). The members of Asian Avenue also took on "the powers that be" at MSNBC during the 1998 Winter Olympics, when the story detailing Tara Lipinski's victory over Michelle Kwan for the gold medal in figure skating ran under the headline "American Beats Out Kwan." The implication that Lipinski (who is white) was an American and Kwan (of Asian descent) was not was thought to be offensive by several members of the EOC. Their subsequent protests eventually elicited an online apology by MSNBC (Ramirez 1998).

One particular example highlights both the amazing potential and inherent flaws of the EOC. In September 1999, one of the headlines of CNETnews.com read:

> Bowing to pressure from the largest Asian American online community, Skyy Spirits is pulling a national advertisement that many mem-

> bers called offensive and degrading to Asian American women. (Girard 1999)

Entitled "Inner Peace," the ad featured a young Asian woman on her knees with a Mandarin dress and chopsticks in her hair pouring vodka for a white woman lying on her stomach clad in only a towel. The ad was debated back and forth in a Politics forum, drawing over three hundred postings. Most AsianAvenue residents felt that the ad reinforced stereotypes of Asian women as subservient. One member referred to the age-old imagery of "china dolls" (Kane 1999). Another complained that advertisements only featured Asians as "spies and hookers" (Girard 1999). Some members were not at all offended by the ad; one responded, "I really think people need to relax. I'm an avid fashion photographer, and I appreciate the picture for what it is" (Girard 1999). The majority of postings were overwhelmingly negative; they were forwarded to Skyy with an accompanying letter from an Asian Avenue.com editor, explaining, "Essentially, Skyy believes that the imagery will sell to a public that buys into these stereotypes" (Girard 1999).

Skyy's marketing director responded with a letter directly to the Web site, stating:

> The intention was to show a woman having a moment with Skyy. A young woman traveling through Asia, who stopped in a peaceful spa to receive a massage in the native tradition of that culture. Thus, the person serving the drink is Asian, i.e., indigenous to the country. Your interpretation of racial stereotyping is distressing. (Girard 1999)

The letter proceeded to apologize, "We are extremely disturbed with the way this ad has raised concerns. Although our intentions were never to do so, the unfortunate outcome is that we have caused offense" (Kane 1999). In a press conference a few days later, a Skyy spokeswoman announced that the ad had been pulled from the campaign and remarked, "We are supportive of the Asian community" (Girard 1999).

Both AsianAvenue.com executives and members seized the opportunity to champion ethnic online communities. Their remarks, however, were tellingly different. CNETnews.com quoted AsianAvenue.com members describing the incident as illustrative of the "potential power Web communities have to unite behind a common cause" (Girard 1999). They also referred to the Web protest as a "coup" for members who complain that their community too often is ignored or stereotyped by marketers" (Girard 1999).

Meanwhile, a Community Connect Inc. executive espoused the protest's "positive" message as a learning experience:

> The lesson isn't that they put out a bad ad. The lesson is that a bunch of people of a similar background discussed how they felt about this ad and how they responded to it. The power of the Web is in polling public opinion, as you couldn't do otherwise. (Girard 1999)

Whether intentional or not, the dichotomy of these EOCs continues to be reinforced. Apparently, the power of community is in shared response, and the power of the Web is in polling. Commodity and community share equal time in the spotlight. Members are excited specifically about their particular community rising above stereotypical interpretation by others. Conversely, the executive's response is much more general and in far more "marketable" terms. The "power" of the EOC must be greater than the ability to poll. The Web enabled the community to extend beyond "a bunch of people of a similar background"; the EOC effected real change in a society that often either ignores or defines minority communities without input from members of said communities. Perhaps Community Connect Inc. and AsianAvenue.com might consider sharing these successes and organizational strategies with its sister Web sites, rather than a Web site mold to be filled in indiscriminately. Such resources might prove more useful in strengthening communities both online and off.

This support is essential because the EOC—despite its intent or worthiness—often represents the particular offline community for which it is named. The media's interpretation of the EOC is insightful, as is Skyy's. When mentioning the different members that contributed to the ad postings, CNET chooses to mention a Harvard professor, an Asian American author, and a New York bar owner (Girard 1999). Apparently, these individuals give the online community respectability—their voices are significant and should be heard. The three hundred postings that comprised the Skyy forum are actually a small percentage of the million-plus membership claimed by AsianAvenue.com. Apparently, the same problems that have plagued society at large and limited minority communities can occur within EOCs as well—decisions for many made by few with privilege. EOCs have the potential to learn from the past and instead empower communities in methods never before implemented. However, a model focused on profit will include only the privileged. Even if the results achieved benefit the community as a whole, the underserved community residents have been excluded from the process of empowerment. Without orientation to and participation in that process, those underserved will always lack the means to rise above their designated status and empower themselves. A community based in technology is limited by lack of access and cannot reach that lofty goal.

Skyy's press conference emphasizes the responsibility that EOCs must bear. The simple statement—"We support the Asian community"—places the weight of the world on AsianAvenue.com. By accepting the mantle and promoting AsianAvenue.com as the voice of the Asian community, Community Connect Inc. must represent the entire community in an effort to bridge differences both within the ethnic group and outside as well.

This is essential; scholars such as Fernback and Thompson (1995) are concerned that online communities may, in fact, discourage interaction with society. They predict that "virtual communities will be communities of interest rather than of geographical proximity or of historical or ethnic origin":

> Just as multiculturalism can and does have a positive influence on self and group identity but when taken to an extreme can disrupt the larger society, so virtual communities can foster anomie. Instead of mass society leading to atomized individuals, however, it may be that it leads to "atomized" communities. And, again, as with multiculturalism, different views can be assimilated into the larger society. But because virtual communities are likely to be private communities of interest, they will not readily or serendipitously be exposed to differ-

ing views that will help them and the larger society grow and adapt to a changing world. (Fernback and Thompson 1995)

This statement reflects a majority view. Ethnic communities are constantly exposed to at least one differing view—that of the majority—because it encompasses the world in which they must coexist. The minority community spends a great deal of its time "adapting" to a changing world—on or offline. However, the Skyy example illustrates the possibility for ethnic communities to actually engage in a societal discourse and expose views that the majority chooses to ignore or is simply unaware of.

As a concluding thought, Fernback and Thompson (1995) add, "If, on the other hand, virtual communities can lead to action, that may be the basis for the formation of real and lasting communities of interest." These are the goals that EOCs must strive to attain. Ethnic online communities become communities of interest when the EOC is interested in the community rather than a financial return. The shared interest of self-empowerment permeates the entire community. That interest then extends outward to ensure that the community remains empowered by controlling and shaping its discourses with other communities and society as a whole. Purpose, not profit, must be the ultimate goal—the stakes are simply too high.

Acknowledgments

Thanks to Dr. David Silver for insight in helping me start, and to Ambika for inspiration in helping me finish.

References

Armstrong, Arthur, and Hagel, John. 1997. *Net.Gain: Expanding Markets through Virtual Communities*. Cambridge, MA: Harvard Business School Press.

Brown, Janelle. 1999. "There Goes the Neighborhood." Online. www.salon.com/21st/feature/1999/01/cov_19feature.html.

Burkhalter, Byron. 1999. "Reading Race Online: Discovering Racial Identity in Usenet Discussions," pp. 60–75 in *Communities in Cyberspace*, ed. Marc Smith and Peter Kollock. New York: Routledge.

Cashel, Jim. 1999. "How To Sell Your Online Community." Online. www.onlinecommunityreport.com/features/subscribe.html (April 30).

——. 2000. "Time for Subscribers." Online. www.onlinecommunityreport.com/features/sell.html (December 1).

"Casting Wider Net, Ethnic Web Firm to Target Blacks and Latinos." 1999, February 9. *Daily News*.

Children's Partnership. 2000. "Online Content for Low Income and Underserved Americans." Online. www.childrenspartnership.org/pub/low_income.

"Economic Viability of Web Forums." 1997. Online. Availabe at www.online communityreport.com/issue.php?date=1997-09-30 (September 30).

Fernback, Jan and Thompson, Brad. 1995. "Virtual Communities: Abort, Retry, Failure?" Online. www.well.com/user/hlr/texts/VCcivil.html.

Gach, Gary. 1998. "Changing Channels." Online. www.asianweek.com/031298/coverstory.html (March 12).

Girard, Kim. 1999. "Vodka maker clips ad after community criticism." Online. http://news.com.com/2100-1017-230488.html (August 31).

Hudson, Michael, ed. 1996. *Merchants of Misery: How Corporate America Profits from Poverty*. Monroe, ME: Common Courage Press.

Kane, Courtney. 1999, September 2. "Vodka Ad Is Stopped After Racism Protest." *New York Times*, C7.

Kolko, Beth. 2000. "Erasing @race: Going White in the (Inter)Face," pp. 213–32 in *Race in Cyberspace*, ed. Beth Kolko, Lisa Nakamura, and Gilbert Rodman. New York: Routledge.

Lee, Edmund. 1999. "Digable Planet Community." Online. www.villagevoice.com/issues/9935/lee2.php (September 1).

Lockard, Joseph. 1996. "Progressive Politics, Electronic Individualism and the Myth of the Virtual Community," pp. 220–30 in *Internet Culture*, ed. David Porter. New York: Routledge.

McClellan, J. 1994, February 13. "Netsurfers." *The Observer*.

Nakamura, Lisa. 1999. "Race in/for Cyberspace: Identity Tourism and Racial Passing on the Internet." Online. www.hnet.uci.edu/mposter/syllabi/readings/nakamura.html.

Ramirez, Anthony. 1998, April 12. "One Site, 14 Ethnic Groups." *New York Times*, 14: 6.

"Reality Bytes." 2000. Online. www.communityconnect.com/articles/wallstreet_062600.html (June 26).

Rheingold, Howard. 1993. *The Virtual Community: Homesteading on the Electronic Frontier*. Reading, MA: Addison-Wesley.

Shafer, Dan. 1999. "Measuring Online Communities." Online. www.online communityreport.com/features/metrics.html (September 30).

Silver, David. 2000. "Margins in the Wires: Looking for Race, Gender and Sexuality in the Blacksburg Electronic Village," pp. 133–50 in *Race in*

Cyberspace, ed. Beth Kolko, Lisa Nakamura, and Gilbert Rodman. New York: Routledge.

Sterne, Jonathan. 2000. "The Computer Race Goes to Class: How Computers in Schools Helped Shape the Racial Topography of the Internet," pp. 191–212 in *Race in Cyberspace*, ed. Beth Kolko, Lisa Nakamura, and Gilbert Rodman. New York: Routledge.

"Ten Business Plans to Watch in 1999." 1998. Online. AtNewYork.com (December 23).

U.S. Department of Commerce, National Telecommunications and Information Administration. 2000. *Falling through the Net: Toward Digital Inclusion*. Online. www.ntia.doc.gov/ntiahome/digitaldivide.htm.

——. 2002. *A Nation Online: How Americans Are Expanding Their Use of the Internet*. Online. www.ntia.doc.gov/ntiahome/dn/anationonline2.pdf.

"A Virtual Enclave." 1998, March 30. *New York Post*.

Wellman, Barry, and Gulia, Milena. 1999. "Virtual Communities as Communities: Net Surfers Don't Ride Alone," pp. 167–94 *Communities in Cyberspace*, ed. Marc Smith and Peter Kollock. New York: Routledge.

"Wired Like Me." 1999. Online. www.communityconnect.com/articles/village_040699.htm (April 6).

Yang, Catherine. 1999. "The Great Equalizer? Not by a Long Shot." Online. www.businessweek.com/1999/99_39/b3648054.htm (September 27).

11

Gay Media, Inc.

Media Structures, the New Gay Conglomerates, and Collective Sexual Identities

Joshua Gamson

In February of 2001, Henry Scott, the former publisher of the American gay-lifestyle magazine *Out*, wrote a letter to some two hundred lesbian and gay activists calling upon them to "help halt an effort to create a dangerous monopoly among gay media." Decrying the recent trend towards mergers in U.S. gay media—notably the purchase of *Out* by Liberation Publications, Inc. (LPI), owner of *The Advocate* and Alyson Books, among other properties—Scott was particularly incensed by the merger of PlanetOut and Gay.com, the largest Web portals aimed at lesbians and gays. "The chief reason for alarm," he wrote, "is that this combination threatens to further diminish the opportunity for vigorous debate over issues of politics and culture and style that is our community's greatest strength. . . . Monopolies don't foster debate, much less creativity and ingenuity" (Scott 2001; see also Bronski 2001). Scott was not alone. "It seems a game of high-stakes three dimensional chess is being played and the pawns are LGBT people whose information sources are increasingly vulnerable to manipulations," wrote New York activist Bill Dobbs, for instance. "Heightening this vulnerability is that I know of no regular platform for self-criticism within the LGBT media" (personal communication 2001). From the other side of the political spectrum, writer Andrew Sullivan complained that the "deeply depressing" consolidation of gay media "into one huge blob" has created a "chilling liberal monopoly at PlanetOut," with "virtually no independent or conservative voices in the mix" (Fost 2001).

The controversy is striking and instructive: It raises important questions about concentration in non-mainstream media and about the impact of the Internet on sexuality-based politics. The very idea that gay media are big enough that charges of monopoly do not sound like the ravings of crazy people speaks to their dramatic growth since the 1990s; until recently, "gay media" called to mind "bar rags," or shoestring local papers, or lesbian-feminist newsletters (Streitmatter 1995), not a multimedia, multimillion-dollar, twenty-four-hour-a-day, goods-and-services- and information-and-interaction-providing conglomerate. It is, to begin with, clear that gay and lesbian media have reached a new historical stage, in terms of both growth and ownership concentration. The early 2000s saw a dramatic trend toward mergers in U.S. gay media: In addition to the LPI purchases in 2000, LPI and PlanetOut attempted to merge (the deal broke down after the .com crash); in 2001, Window Media, owner of the Atlanta-based *Southern Voice*, *Houston Voice*, *Eclipse* magazine, and the Atlanta lifestyle quarterly *SOVO*, acquired the *Washington Blade* and the *New York Blade News*, giving it control of several of the most important gay and lesbian regional publications (Stern 2001); 2001 also brought the Gay.com merger with PlanetOut. Gay media are now very much modeled on their mainstream counterparts, especially in the trend toward concentrated ownership; when PlanetOut announced plans to acquire Liberation Publications, CEO Megan Smith liked to say the deal was "the gay and lesbian version of the Time Warner-AOL merger" ("Gay Web Portal" 2000).

Mainstream media conglomeration has in fact received plenty of scholarly attention. In 1983, Ben Bagdikian famously wrote with alarm of "the media monopoly," in which about fifty firms controlled more than half the global media; with each subsequent edition of his book, the number of controlling firms diminished through ongoing series of mergers and acquisitions: from fifty in 1983 to twenty-three in 1990, to ten in 1997; now, with the AOL-Time Warner merger, the number has dropped to eight (Bagdikian 1983, 2000). The concentration of media ownership is accompanied by both vertical and horizontal integration of the industry. A single conglomerate such as Disney or Viacom, in addition to being linked to nonmedia industries through ownership or interlocking board directorates, owns mechanisms of both production and distribution and has holdings across a variety of different types of media (Compaine and Gomery 2000; Croteau and Hoynes 2001).

Most theorists are alarmed by this, either because it represents reduced competition or because it represents increased hegemonic control; critical media studies scholars assert that ownership consolidation decreases content diversity (of demographic groups represented, points of view, editorial positions), bringing it into line with the interests of a small number of corporate owners, squeezing out independent companies and alternative perspectives, and generating corporate homogeneity (Croteau and Hoynes 2001). Concentration, it is asserted, brings both an enormously increased interest in suppressing critical information and voices and in the capacity to suppress them; the fewer and larger the number of owners, this reasoning goes, the more likely it is that radical, critical voices, already edged out by commercial pressures (Bennett 1996), will be pushed off of the media map. As Douglas Kellner puts it, "In a capitalist society where the means of communication are concentrated in powerful corporations, the access of minority, oppositional, or alternative views is denied or limited" (Kellner 1990, 94).

Although such concerns are understandably imported by critics of gay and lesbian media ownership concentration, the dynamics and impact of ownership concentration in the media systems of marginalized or minority groups are not well understood. Even if we accept assertion that ownership concentration reduces the diversity of media content—which remains open to debate in the field (Croteau and Hoynes 2001; Napoli 1999)—there is no reason to accept the assumption that non-mainstream media operate just like their mainstream counterparts. After all, minority media systems emerged from social movements, and they have played particular roles in particular political struggles. The gay and lesbian "monopoly" controversy points out how much remains to be understood about the dynamics and impact of ownership concentration in the media of marginalized groups.

Moreover, the alarmed discussions about *new* gay media—after all, it is Internet companies that are the trigger for the alarm—also pose a remarkable contrast and an important challenge to existing scholarship on computer-assisted communications media in general, and on sexualities and new media in particular. Even if Scott and other critics are wrong in their monopoly charges—and, at least for now, the charges are overstated at best—they point to issues that have been virtually ignored in critical scholarship on new communications media, especially under the cultural studies rubric. Scholars have sobered up a bit recently, but theory and research on new media have nonetheless been much more interested in agency than structures, and in discontinuities than continuities: new forms of individual identity, new virtual-community formations, new forms of representation (Gauntlett 2000; Hillis 1999; Jones 1997; Turkle 1997). As Korinna Patelis, writing in the late 1990s, described it, a sort of "Internet-philia" has often characterized the field, characterized by the announcement of "the inevitable arrival of a whole new era" and "a clear break with the past," brought on by the shift from analogue to digital information, which "enhances freedom" and "removes the limitations of geography," is "decentralized . . . and thus cannot be technically controlled," and is "anti-statist." The vision was one in which "no structures will prevail, no knowledge will be diachronic, no policy definite, no question permanent" (Patelis 1999).

Thus, in the literature on new media and queer sexualities in particular—perhaps because sexuality is so tied to the body, and nonconforming sexual identities and communities so self-consciously generated—one hears repeatedly of the queerness of cyberspace: how, as David Gauntlett summarizes it, "the Internet's scope for anonymous interaction, and therefore identity play, is significant for the way in which it fits in with contemporary queer theory," bringing the notion of identity as fluid and performative to life by breaking "the connection between outward expressions of identity and the physical body which (in the real world) makes those expressions" (Gauntlett 2000), and how new technologies have provided a breeding ground, if you will, for "the evolving cybercultures of sexual dissidents" (Bell and Kennedy 2000, 392), or what Nina Wakeford calls "cyberqueer spaces" (Wakeford 2000, 408). These approaches have yielded important observations, and they are clearly not without merit, since computer-assisted communications *are* relatively unstructured and decentered and evasive of control, and in many ways people *are* actually using them for generating and trying on versions of the self and community that defy conventional locations in the body and geography; sexual cyberculture and identity play are terrific examples. Still, there seem to be big chunks, theoretically and empirically, that remain overlooked.

As the pendulum swings back toward what one analyst, Andrew Shapiro, calls "a kind of 'technorealism'" (Stille 2001), government and corporate control (McChesney 2000) and the downsides of personalization and social fragmentation (Sunstein 2001) are becoming analytically central. Yet the legacy of Internet-philia remains: most notably, in the tendency to bracket institutions and ownership, to research and theorize uses and users of new media outside of those brackets, and to let "newness" overshadow historical continuity. That strategy is not just incomplete, but risks distortions: overstating individual and collective agency can obscure important structural changes the Internet has brought, and leaves unexamined of the link between those changes and the shape of sexual identities, communities, and politics; overemphasizing newness can hide significant, if unglamorous, long-standing processes being advanced by new technologies. In the discussion that follows, I put those on the table, as a centerpiece, and to recommend that organizational and institutional analysis—how new media organizations operate, the kinds of institutional structures being built, their economic and organizational logics (DiMaggio et al. 2001)—be a continuous and firm part of understanding sexualities and new media. (This is not a new suggestion, but one worth repeating until it takes.)

In a modest step here, I consider the story of PlanetOut and Gay.com, now PlanetOut Partners, Inc., and what it can begin to tell us and ask us. The PlanetOut–Gay.com merger is useful in directing our attention to structures of ownership and control, rather than the playful, liberatory, evasive uses (and users) of computer-assisted communications; to the ways new media influence already-existing conservative trends in sexual identity politics, consolidating political elites and building solidarity-in-consumption, rather than the ways they undermine conventional sexual identity politics and "queer" them up; to the symbiotic relationship between new gay media and old gay media institutions, rather than the relationship between users and new technologies. It is especially useful in challenging us to rethink how commercial enterprises, technologies, and collective identity boundaries intersect.

At a general level, my argument is that everything new is old again. More specifically, I will suggest that, at least in the United States, exactly *because* of the new identity possibilities, virtual relationships, and geographical independence it allows, the Internet has been a major force in expediting, amplifying, and solidifying historical processes that began to take hold in the 1980s: the transformation of gay and lesbian media from organizations answering at least partly to geographical and political communities into businesses answering primarily to advertisers and investors; the consolidation of lesbian, gay, bisexual, and transgendered media ownership; and a tightening of relationships between mainstream political organizations, non-gay corporations, and national gay and lesbian media. In this case, it seems that media market concentration is not leading in any unambiguous way to the narrowing or stifling of voices; its unambiguous impact is in the final decoupling of media from movement. The closing-in, closing-up, and closing-off processes are, ironically, brought to us by the famously opening and freeing Internet.

Venture Capital and the Structural Features of Internet-Based Conglomerates

PlanetOut and Gay.com were part of the growth in the 1990s of Web portals, which serve as points-of-entry to and gatekeepers of the Internet, organizing attention and classifying information in an environment comprised of billions of Web pages and extreme "information abundance" (Hargittai 2000). Quite quickly, major Web portals became the destination for a huge proportion of Web traffic. Gay portal companies also capitalized on the apparently unusual appeal of Internet to gays, lesbians, bisexuals, and transgendered people (Meland 2001). As a result, new communications media have facilitated the largest investment of capital in American gay and lesbian businesses *ever*. In 1995, PlanetOut became the first company targeting the gay market to secure venture capital, and in 1999, the company raised $16.4 million in second-round funding, from America Online Investments (which owns 12 percent of the company), the Mayfield Fund, and Eden Capital, as well as from private investors such as Excite senior vice president Joe Kraus, E*TRADE president Kathy Levinson, Nicholas Negroponte of MIT's Media Lab, and Real Networks CEO Rob Glaser (PlanetOut 2001a). In a third round of funding in 2000, the company raised another $10 million, this time adding BMG Entertainment, the Creative Artists Agency, and EDventure Holdings (whose majority owner is techworld goddess Esther Tyson) to the list of investors (PlanetOut 2001d). PlanetOut has distribution partnerships with, among others, AOL, Netscape, Yahoo!, Snap.com, Lycos, Real Networks, and CompuServe. Its advertisers include Arista Records, Atlantic Records, Barnes and Noble.com, Coors Brewing Company, Fox Searchlight, IBM, Johnson and Johnson, Levi Strauss, Macy's, MGM, Procter and Gamble, Saturn, SmithKline Beecham, Sony, Starbucks, United Airlines, Virgin Atlantic, Wells Fargo, and many others (PlanetOut 2001b).

Until recently, PlanetOut's major competitor was Gay.com, whose parent company, Online Partners, in 2000 raised $23 million in venture capital, with investments from Chase Capital Partners, Flatiron Partners, Baroda Ventures, and IDG Ventures, among others. It hosts many of the same advertisers and sponsors, such as IBM, Johnson and Johnson, and Saturn, and also includes American Airlines, AT&T, Gap/Banana Republic, Glaxo-Wellcome, Pharmacia and Upjohn, Propecia, Saturn, and Time (Online Partners 2001a). Such massive, sudden investment of money in these companies was due, of course, in part to the gold-rush mentality of the .com boom, and after the fall investors are harder to come by. But it is undeniably significant that the most dramatic investment of money into gay and lesbian businesses took place on the Web.

Gay.com and PlanetOut quickly began to tie themselves, through deals and acquisitions, to other companies, driven by general business logic and the specific exigencies of the Internet—hence the second major structural feature of the companies, conglomeration. For Internet startups in particular, the main difficulties are establishing "brand" recognition and generating "content." PlanetOut and Gay.com took a logical short-cut: buying or trading with existing properties, especially older, established media companies, which provided content, legitimacy, and "brand" recognition. Gay.com developed strategic alliances with the *New York Times*'s Internet division and the Gay Financial Network and content partnerships with gay and lesbian publications such as *The Blade* and *Girlfriends*; the company has acquired the Gaywire News Network and the online service OnQ and has teamed up with the independently produced national gay and lesbian television series "In The Life." In 2000, PlanetOut bought the travel newsletter *Out and About* and the digital rights to the gay youth magazine *XY*; it has partnerships with Culturefinder.com, Canada's City TV, Kleptomania.com, and the Ifilm Network; although its deal to acquire Liberation Publications fell through, *The Advocate* remains a partner, and the incentive to eventually merge the companies remains. And, of course, PlanetOut and Gay.com, friendly but fierce competitors in their early years, have now joined forces to create what the former board chairman of the new entity, PlanetOut Partners, called "a powerhouse media company with tremendous reach within the lesbian and gay community" (PlanetOut 2000), and what critics see as a monstrous "monolith" (Bronski 2001). In its new incarnation, in addition to laying off employees, PlanetOut Partners quickly attracted another $8.2 million in capital investments, and several high-profile new advertisers (PlanetOut 2001c).

The staffing of these companies underlines, finally, just where PlanetOut and Gay.com are institutionally located. Those running PlanetOut Partners, unlike many of those in older print media, emerged from the business world rather than from anything remotely like activism. The official biography of Megan Smith, now president of PlanetOut Partners, emphasizes her master's degree in mechanical engineering from M.I.T. and her work at Apple Computer in Japan and then at General Magic, Inc., where she "closed $9 million in equity financing from Toshiba and Nortel," among other achievements (PlanetOut 2001a). Lowell Selvin began with the high-tech company Light Signatures, and spent eighteen years "in rapid growth, early-stage and turnaround business management and consulting," with clients ranging "from start-ups to Fortune 500 companies, Johnson and Johnson to Hilton, MGM to Electro Rent." He cofounded Degree Baby Products, which was acquired by Johnson & Johnson and served as CEO for Arbonne International, a direct sales and marketing company, and as a director for Arthur Andersen Business Consulting (Online Partners 2001a). The company's Board of Directors draws almost entirely from its venture capital funders: the managing partner of Flatiron Partners, a general partner from the Mayfield Fund, a vice president from AOL-Time Warner Ventures, and so on. They believe in a community mission, as well, but not surprisingly those high up in the companies resist any notion that they are *accountable* to "the community," which is neither their primary network nor their primary concern. "People want to treat us as a nonprofit organization representing the community," says PlanetOut, Inc.'s executive vice president Susan Schuman, who for a time was managing director of the Human Rights Campaign. "And we're not. We're a business, and we provide goods and services." This is a significant structural development in the gay, lesbian, bisexual, and transgender media world: a single dominant company, merging with and acquiring media properties old and new, controlled by a single board of directors, owned by and answering to a handful of corporate and individual investors, aiming to be ground zero of the gay and lesbian cultural and informational system.

The Internet and Marketing-as-Liberation Rhetoric

How have these companies managed to convince investors and major advertisers, even after the collapse of the .com market, that theirs was the media business that would pay off? Although PlanetOut Partners, structurally speaking, really does look like mainstream media company, the company's marketing strategies are more specific to minority media—and interestingly, they build specifically on arguments about the Internet as a hotspot for sexual minorities. Not surprisingly, the primary strategy has been to bring to fruition the kind of marketing-as-liberation logic that has characterized much of gay public life for the last couple of decades, documented in exhaustive, smart detail in Alexandra Chasin's *Selling Out* (Chasin 2000). This reasoning has nothing in particular to do with *new* media, but simply extends the familiar arguments made, with limited success, by print media publishers looking to attract funds. Gays and lesbians are a "relatively untapped market" (PlanetOut CEO Megan Smith) and an "attractive market segment" (AOL Investments Vice President Ron Peele) that is brand loyal, with disproportionate buying power (PlanetOut 2001d). "Gays and lesbians show a preference for products advertised to them an amazing 84% of the time," says PlanetOut's pitch to advertisers. "16 million loyal customers with $450 billion in buying power are ready and waiting" (PlanetOut 2001b).

The folks at PlanetOut and Gay.com, like those in many other businesses targeting marginalized groups, like to dance at the wedding of doing well and doing good, and there is no reason to doubt their sincerity. Even PlanetOut's business cards display the equation of marketing and liberation. Each has a detachable flap with a factoid on the opposite side. It might quote a question to one of the Web site's advice columnists about the telltale signs that your child is gay, or it might tell you that 20 percent of gay and lesbian high school students will be injured by their peers, or that two-thirds of guidance counselors harbor negative feelings toward gay and lesbians. Or it might tell you how favorably gay people's buying habits compare to the "general population": three times more likely to have a money market or mutual fund; six times more likely to take six or more flights per year; four times more likely to have an American Express card. Lowell Selvin, Gay.com's CEO, wrote this in a recent column on the site of the company's "new business model as a multi-mission organization":

> Our company must serve the community in order to prosper. For us, it's good business to be altruistic. . . . So, this humane, community-based mission, how does it impact our economic one? What do the venture capitalists say? You're in for a surprise. All of our major backers, whether straight or gay, get that we need to serve the GLBT community with passion, human and financial resources, and the strength of our global voice unparalleled in human history. It's good business: Our investors and leaders know, as with any product, that our business must appeal to, and engage the support of, the people we want to do business with. (Selvin 2000)

Similarly, PlanetOut CEO Megan Smith captured the ideology of marketing as a mechanism of political change. When a company advertises to lesbian and gay people directly, she said,

> One, you're reaching a person and you're telling them your message, which is true anywhere, but second, you tell them that you, Procter & Gamble, or you, Saturn cars, or you, Saab, don't have an issue with them and who they are. So you say, not only do I have great cars, but I also think that you're great, so go buy my car. (author interview 2000)

Serving the community and penetrating the market are one and the same. Those running these companies argue that, as Alexandra Chasin puts it, "their own entrance into the marketplace [is] a corrective for past social alienation" and that "gay identity-based consumption [will] ameliorate homophobia" (Chasin 2000, 39).

Again, there is nothing new about this market-niching of gay and lesbian lives, or of other minorities (Davila 2001); what's new with PlanetOut and Gay.com is the degree of success they've had in getting others to buy into the strategy. This success has largely to do with the additional arguments the companies add to their pitch, which are specifically related to *new* technological capacities that make the Internet an especially gay-friendly space. The first is simply that the Internet allows them greater "reach" than ever before. Gay.com is an international network, with Brazilian, Italian, British, Mexican, French, and Argentine branches. The companies claim that there are an estimated 17 million gays and lesbians, and that 10 million of them are online; each company claims over 1 million registered users, and combined, PlanetOut and Gay.com "claim to have 4 million unique visitors a month, or 35 percent of the online gay market" (Cohen 2001). Because of new technology, they are able to reach—and sell to advertisers—exponentially more people than all gay and lesbian print publications combined, and reliably hit their gay and lesbian target more directly than, say, an ad on NBC during *Will and Grace*.

More interestingly, however, these commercial sites are built on an argument—echoing academic writings, although in the language of marketers—about *why* the Internet is such a great place for gay and lesbian people to congregate, and therefore for advertisers to target them. They rely for their success, ironically, on the strength of the closet. The closet inhibits both community-building and market penetration, and the Internet, by allowing people to consume, observe, learn, and interact regardless of geographical location and by offering the possibility of anonymity, opens them back up. “At a fundamental level, what PlanetOut does is it *reduces isolation*,” PlanetOut CEO Megan Smith said. “Then layered on top of that, it entertains and informs and provides goods and services.” About two-thirds of what happens there, according to Smith, is “what our customers do with each other” (personals, polls, message-board discussions, e-mails, chat) (author interview 2000). “At the end of the day,” says PlanetOut Executive Vice President and General Manager Susan Schuman, “it's about people wanting to find people like them” (author interview 2000). Indeed, Gay.com established itself largely through the chatting and cruising and cybersexing that its customers do with each other in chat rooms, and built from there; clearly, although it can be very risky and scary face-to-face, online cruising has almost no risks whatsoever. PlanetOut draws its customers primarily by offering them what its promotional materials call a “vibrant, welcoming and safe community”; once there, maybe they'll shop, and even if not, they can certainly look at ads.

While one can certainly take issue with the use of the word “community” to describe such mall-like environments, it seems indisputable that the Internet, with its wacky-screen-names anonymity and freewheeling sexuality, is indeed a particularly great place for those people, especially those who are isolated, who do not want to out themselves, who are uncertain in their sexual desires, or who want to test the waters, meet potential partners, hang out with other gay people, or get information about gay or lesbian or bisexual or transgender life without anyone necessarily knowing they're doing it. The argument that this is a huge business opportunity is thus compelling. “If you can get fired in 39 states [for being gay], you're not going to walk up to a newsstand and pick up a gay magazine, and therefore the newsstand doesn't carry the magazine,” says Smith. Even *Out* is not out: It's delivered in a plain white envelope. “The Internet breaks that problem” (author interview 2000). And for those who are openly gay, lesbian, bisexual, or transgendered, the Internet is a resource for finding people and information that the print and broadcast media, and a don't-ask-don't-tell culture more generally, make difficult to find. Joining an online conversation without worries of repercussion from family or employers, consuming information and services that offline would mark you as gay and put you at risk, getting off with someone apparently of the same sex, "lurking" unseen and unheard, trying on identities without commitment, transcending geographical location to discuss issues and organize—all these activities are exactly the kinds of things that many scholars have seen as generating a cyberworld of fluid identities, queer selves, and decentered communities. They are also the key to the bank.

So What? The Significance of Internet-Based Gay Media Conglomeration

The picture painted thus far ought to give pause to Internet-philes, especially those touting new media as primarily a force of opening for sexual communities and identities. The notion that the Internet works against central structures is at best incomplete, at worst dangerously obscuring; far from fragmenting and decentering control of gay media and culture, the Internet here has concentrated it and facilitated the growth of the first gay media conglomerate: an unapologetic mini-me of its massive, merger-loving counterparts, sponsored by unprecedented investment of venture capital, successfully becoming a powerful center not just online but offline as well. The celebrated new-media features—of "anonymous interaction, and therefore identity play," of a severed "connection between outward expressions of identity and the physical body" (Gauntlett 2000), of "virtual community" cyberspaces that combine "the connected sociality of public space with the anonymity of the closet" (Woodland 2000, 418), of "on-line queerspaces" where "identities are shaped, tested, and transformed, both individually and corporately" (Woodland 2000, 430)—are, for PlanetOut and Gay.com, an opportunity to build and pitch their "new business model as a multi-mission organization," to create a central "destination" for millions of people, and to attract unprecedented cash. And, as I will later suggest, the version of collective identity they articulate in the process, while it is certainly pluralistic, seems hardly fluid, especially queer or post-anything.

There is also much to be gleaned from the PlanetOut-Gay.com story for those interested in and concerned about media market concentration. In particular, it offers an opportunity to consider whether and how and when ownership concentration results in narrower and more homogenized content. Such a process, this case suggests, is not as simple or direct as critics tend to assume. For one thing, the dynamics vary by medium. On the Internet itself, where there is genuine, persistent, unstoppable competition for attention, there are some real limits to monopoly: Media conglomerates cannot stop computer-based sites from popping up and from showing up very easily for surfers. (This limitation should not be overstated, however: Search engines and mainstream portals "systematically exclude [in some cases by design and in some accidentally] certain sites in favor of others," so that "content produced by entities with large enough budgets can attain prominent placement," privileging their "content over material by smaller and less financially-endowed creators" [Hargittai 2000, 9, 23].)

More importantly, PlanetOut and Gay.com operate according to different logic than older broadcast and print media: The same business interests that drive mergers and acquisitions lead the sites in many ways to promote rather than suppress the raggedy diversity—especially cultural and demographic diversity—of lesbian, gay, bisexual, and transgender populations. Such "affinity portals" cast their net broadly; their popularity is measured not just in registered users but in "unique visitors," in "traffic" of attention. They primarily want to get you to stop by the site habitually, to use them as your point of entry, even if you move from there to other linked sites (which itself generates "click-through" revenue). The more you see their sites as a gateway into all things GLBT, the more successful the company, and linking to and thereby promoting (rather than suppressing) smaller sites of all kinds only increases the chance that PlanetOut or Gay.com will be your first stop. In fact, that is one of the main reasons that they eat up, and then maintain as-is, smaller sites. In these ways, it is no lie to say that content diversity is good business. Much of the sites' content (personals, discussion and chat groups, etc.) is generated, as well, not by editors or producers but by interacting members—owner control of content is considerably looser than in print or broadcast media. Finally, on the Internet, unlike in print or broadcasting media, it is relatively easy and inexpensive to be inclusive; there is little financial cost to casting a wide net. Unlike in older media, where adding a page or a minute requires finding advertising revenue to pay for it, "we don't have to make the decision that a print publication does of where to cut off the information," PlanetOut Programming Manager Matt Alsdorf remarked, since new pages and links can be added at relatively low cost (author interview 2000). Content diversity serves the interests of the business and is cheap. That's all business logic; any moral commitment to "diversity" is gravy.

These sites have only a limited interest in producing original news (Smith 2002); although it is very difficult to measure Internet content over time, the new media conglomerates do not devote a lot of resources to their news divisions, and as they consolidate the news their subsidiary companies produce does appear to homogenize, since costs are cut by circulating the same news story in the company's various outlets (Montopoli 2002). But given the Internet's business logic, on the cultural front at least, content is expanded rather than narrowed. If cultural homogenization were the main operating principle here, one would be hard-pressed to explain the presence on these sites of hard-to-find clips like one of anti-gay crusader/orange-juice spokeswoman Anita Bryant getting pied in the face, the Charles Nelson Reilly Jell-O commercial, and a homoerotic Laurel and Hardy episode; columns by transsexual activists and writers Kate Bornstein, Susan Stryker, Jamison Green, and African American gay folks such as Keith Boykin; profiles of Urvashi Vaid, Ruth Ellis, Dorothy Allison, Audre Lorde, and Marlon Riggs; expert advice from Dr. Dyke, leather daddies, and Nina Hartley; or links to "sensual transgender fiction," Dyke Marches, "S/M smut," the Asian Lesbian Bisexual Alliance, the Sexual Freedom Coalition, and the San Francisco Leather Pride group, for starters. PlanetOut and Gay.com must also be presentable to advertisers and investors, and as a result the sex talk, the insistent and complicated mixes of racial and gender and sexual (and sometimes class) identities, the confident perversions—the things that might remind advertisers and investors of the dissident parts of gayness—are pushed below the rainbow-flaggy, advice-from-Ellen's-mother, celebrity-profile, shop-til-you-drop front pages. But unlike in many of its print siblings, those images and voices and ideas are a click away. For all its crass commercialism—indeed, because of its crass commercialism—this is a more pluralistic cultural setting than most LGBT institutions.

The impact of the new gay media conglomerates, then, is not necessarily to decrease the overall diversity of media content and voices—unless of course they go under, taking the old media companies they've bought down with them. (This is what happened to Australia's Satellite Group, a gay media and real estate company that went public and had to shut down seven gay publications when it went into receivership [Bronski 2001]). Their most direct and observable impact is nonetheless extremely important. They are solidifying a shift in social relations among institutional actors that began before them: further strengthening the ties between gay media owners, advertisers, and corporate investors and loosening those between gay media and LGBT political activists.

The gay media and the gay movement have historically been "interdependent and co-emergent" (Chasin 2000, 90; Streitmatter 1995). Now, the PlanetOutish world of media business and the old-style grassroots activist world, which called one another into existence, are increasingly distant. Gay media, that is, have unquestionably shifted location, and the Internet has more or less completed the shift. Literally: When PlanetOut, having outgrown its startup space, went looking for a new one, it moved from the activisty Mission neighborhood into the .commiest San Francisco neighborhood, South of Market—a symbolic move, says Schuman, "in amongst its Internet peers" (author interview 2000).

When activists like New York's Leslie Cagan say, "I want a media, or at least some media, that sees itself, consciously, as part of a social change movement" (author interview 2000), it feels almost like nostalgia. The growth of new gay media conglomerates may simply be the last nail in the coffin of the gay media as movement "property" and movement organs; the primary function now of gay media is quite clearly to deliver a market share to corporations, and the ties fostered by the consolidated gay .coms have firmly institutionalized that change. Other than a link on the Web site, activists and social-change organizations are out of the institutional loop.

The first moves of the new PlanetOut Partners in 2001, besides raising another round of funding and laying off some staff, offer some clues in this regard. One was to announce an alliance with Socratic Research Technologies, Inc., an international market research company specializing in high-tech and media industries, to form "the world's most comprehensive panel of lesbian, gay, bisexual, and transgender people." By enabling "key insights into community needs," the company offers to provide a service to "manufacturers and service providers who wish to understand the LGBT markets better" (Online Partners 2001c); community needs are conflated with consumption desires, and community equated with market. The key relationship being fostered is between PlanetOut Partners, market researchers, and "manufacturers and service providers." The second major announcement was the first-ever gay pop music festival, called Wotapalava Music Festival, a "Lollapalooza-style event geared toward the LGBT community" promoted by the Creative Artists Agency, Inc. (CAA) and headlined by the Pet Shop Boys, of which PlanetOut and Gay.com are "premier media sponsors" and the exclusive sellers of advance tickets (Online Partners 2001d). (The event was canceled after low ticket sales.) Key relationships are being built between PlanetOut Partners, CAA, and music industry sales people, exhibitors, and marketers. On June 5, 2001, the company announced that it had presented the "Gay.com and PlanetOut.com Gayest Commercial of All Time Award," which "honors advertisers that are breaking ground with outstanding commercials that effectively communicate to gay and lesbian customers through gay and lesbian themes," to Leo Burnett Worldwide, Inc. and Hyundai Motor Company for its commercial "Boy Toy" (Online Partners 2001b). The ad, of course, can be viewed at either site. In October 2001, PlanetOut Partners announced a partnership with Rivendell Marketing and several other companies (a mailing list brokerage, an ad agency, a strategic marketing communications firm, a marketing and publicity agency) to form the Gay Media Alliance. The Alliance aims "to create integrated marketing programs for Fortune 500 companies" with "a variety of media and marketing services to create customized programs for branding, product positioning, corporate awareness, new product launches, product promotion and direct response programs" (PlanetOut Partners 2001). Plainly, PlanetOut Partners is establishing and tightening a very particular network of alliances and accountability that has little to do with social movement activists.[1]

New Media, Old Identities

Much of what I've argued is that the growth of Internet-based gay media conglomerates continues, rather than breaks with, the commodification-of-homosexuality trends and the distancing of gay media from LGBT-activist movements that began earlier and elsewhere—another reminder, if we need one, that it is worth discarding the fetish for "newness" that has often characterized the field of new-media theorizing. Their impact, then, has largely to do with their strengthening of the hegemony of gay consumption-as-liberation ideologies, their solidification of the social and institutional ties that support that ideology, and their merging with "old" media—further reminders, if we need them, that it is also worth challenging the lack of structural analysis and the centering of agency that has characterized the field, and especially so much of the gay, queer, and feminist writing on cyberspace.

It is not a matter of trading structure for agency, or control for liberation, in attempts to understand the impact of new technologies on sexualities. What remains ill understood, for instance, is the *relationship* between the institutional dynamics I have recounted and the building of identity. Institutions, for instance, are a major source of the "cultural resources " or "tool-kits" (Swidler 1986; Williams 1995) from which identities are built; it is worth considering what sorts of cultural resources companies like PlanetOut Partners are providing, especially if they succeed in achieving the cultural centrality to which they aspire. The role of new gay media companies in how collective identity—the sense of "we" of a group, of who is and is not a member, of what membership means and what binds members to one another (Poletta and Jasper 2001; Taylor and Whittier 1992)—remains to be seen. One possibility implied by the story I've told is that, as they further the transformation from movement to market, the consolidated gay .coms simply give more power to the notion that the most legitimate members of the LGBT "we" are those with the most money, prestige, and privilege. As Chasin argues, this overall transformation brings with it greater visibility for "gay men and lesbians with money," a sort of "unintended disenfranchisement [on the basis of race, class, and gender] that is an effect of conceiving of political rights as market-based rights"; an articulation of identity that gives sexuality primacy, often thereby ignoring "differences among gay men and lesbians, such as those of gender and race"; and a tendency toward a "nationalist," pseudo-ethnic paradigm, since the group must be a definable and identifiable market segment (Chasin 2000, 7, 20, 92, 45–46). Indeed, in their unapologetic promotion of the gay market niche, companies like PlanetOut Partners seem to have simply engendered, strengthened, and profited from an old class- and race-inflected "minority" model of group identity, and a version of the self that assumes sexuality as primary and essential and unchanging (Epstein 1987; Gamson 1995; Valocchi 1999). New technologies, as they are used to build commercial media institutions, wind up being used in ways that are quite hostile toward queer ideas of fluid, intersectional, performative, dissident, and challenging sexual identities. Folks may, of course, still build their identities from other material, they resist and play and so on, but the terrain has changed, and its growing center is hardly queer.

But it is peculiar. It seems just as much the case that, while ardently promoting this version of an "ethnic-essentialist" sexual minority (Seidman 1993), companies like PlanetOut Partners are loosening collective boundaries in new ways, as profit-seeking enterprises often seem to do (Gamson 1998). The business logic I've recounted, for instance, is not simply to sell rich, white gays and lesbians to advertisers, but to create out of the fragmented, divided, complicated subgroups a sort of mass market, to gather them together in layers; this is a strategy encouraged and made smoother by new technologies. The companies have a financial interest in keeping the boundaries of the "we" *loose* rather than tight, and it costs them very little to do so. They exhibit very little of the sharp-edged boundary-vigilance one typically finds in activist organizations (Gamson 1997), or even in the more organic, noncommercial online gatherings, in which, as Cass Sunstein has recently argued, "individuals bypass general interest intermediaries and restrict themselves to opinions and topics of their own choosing," mainly "listening to louder echoes of their own voices" (Sunstein 2001, 16).

At the very least, theory and research on new media and sexualities, as well as on media ownership concentration, might be well served by beginning from the complicated, ironic tensions that the story of PlanetOut Partners encapsulates: the openings that close, and the closures that open. Technologies celebrated for loosening and decentralizing control of media production, for promoting unconventional identities, become the basis for a tightly controlled, highly conventional gay media conglomerate. The transformation of the lesbian, gay, bisexual, and transgender movement into a market, which requires narrow and palatable versions of identity and community, is taking place through businesses with a particular interest in casting the identity net very wide. The result, oddly, is a faux-community quite uninterested in policing who belongs, a commercialized cyberworld whose group boundaries are looser—not fluid, but relatively inclusive and porous—than those of the movements from which they have so cavalierly and controversially detached.

Acknowledgments

This chapter has benefitted from discussions at colloquia presentations at the University of Surrey, New York University, the University of California at San Diego, and the University of California at Irvine and from commentary by students in my Writing Sociology workshop at Yale University. I am especially grateful to Nina Wakeford, Clare Hemmings, and David Meyer for detailed criticism. The research was supported in part by grants from the Whitney Humanity Center's Griswold Fund and the Fund for Lesbian and Gay Studies at Yale University.

Note

1. The relationships between gay media owners, advertisers, and investors are not the only ones the Internet has brought to the fore. Some critics have suggested that the already-cozy relationship between national U.S. gay magazines such as *The Advocate* and mainstream gay political organizations, in particular the large lobbying group Human Rights Campaign (HRC), has been and will be made even tighter through the new gay media conglomerates, and be even more insulated from criticism. There is little evidence at this point to support or refute the latter charge, but critics point to the problems raised by the organizing of the Millennium March, a large civil rights event organized largely by the HRC. The three largest corporate sponsors were the Advocate/

LPI, Gay.com, and PlanetOut, which alone donated cash and in-kind contributions worth $1 million (Bronski 2001). For a variety of reasons, the march generated considerable controversy and opposition within lesbian and gay political circles–its organizing process was seen by some as exclusive, shady, and corporate-style and its goals as either vague, conservative, or money-driven among gay and lesbian activists (Gamson 2000), very little of which received coverage in the publications of the three major sponsors. Whether this demonstrates the shape of things to come, the overall reasoning is quite clear: Not only do PlanetOut Partners and HRC share a basic philosophy of mainstreaming-through-economic-clout and a history of informal social ties, but PlanetOut Partners (and *The Advocate*) are major sponsors of the work of HRC, among others. That is, to the degree that new gay media are involved in movement politics, it is now primarily through sponsorships and contributions, often to events in which they themselves have invested, and disproportionately to events that will expose the largest number of people to their "brand"; on the flip side, national organizations cannot really afford to be critical of the increasingly consolidated gay media, since they depend on them to publicize their organizations and their work and to help fund them.

References

Alsdorf, Matthew. 2000. Author interview, San Francisco, June 15.

Bagdikian, Ben. 1983. *The Media Monopoly*. Boston: Beacon Press.

——. 2000. *The Media Monopoly*, 6th ed. Boston: Beacon Press.

Bell, David, and Barbara M. Kennedy, ed. 2000. *The Cybercultures Reader*. London: Routledge.

Bennett, W. Lance. 1996. *News: The Politics of Illusion*, 3rd ed. White Plains: Longman Publishers.

Bronski, Michael. 2001, March 1. "Then There Was One." *Boston Phoenix*. [Page numbers unavailable.]

Cagan, Leslie. 2000. Author interview, New York, June 2.

Chasin, Alexandra. 2000. *Selling Out: The Gay and Lesbian Movement Goes to Market*. New York: Palgrave.

Cohen, Jackie. 2001, March 16. "Liberation Publications and PlanetOut End Plans for Merger." *Press Pass Q*. [E-mail newsletter.]

Compaine, Benjamin, and Douglas Gomery, ed. 2000. *Who Owns the Media: Competition and Concentration in the Mass Media Industry*. Mahwah, NJ: Lawrence Erlbaum Associates.

Croteau, David, and William Hoynes. 2001. *The Business of Media*. Thousand Oaks, CA: Pine Forge Press.

Davila, Arlene. 2001. *Latinos, Inc.: The Marketing and Making of a People*. Berkeley: University of California Press.

DiMaggio, Paul, Eszter Hargittai, W. Russell Neuman, and John P. Robinson. 2001. "Social Implications of the Internet." *Annual Review of Sociology* 27: 307–36.

Dobbs, Bill. 2001. Personal communication, March 27.

Epstein, Steven. 1987. "Gay Politics, Ethnic Identity: The Limits of Social Constructionism." *Socialist Review* 17(3–4): 9–54.

Fost, Dan. 2001, February 1. "San Francisco Firm also Adds Web Site, Charges for Net Ads." *San Francisco Chronicle*, B1.

Gamson, Joshua. 1995. "Must Identity Movements Self-Destruct? A Queer Dilemma." *Social Problems* 42(3): 390–407.

——. 1997. "Messages of Exclusion: Gender, Movements, and Symbolic Boundaries." *Gender and Society* 11(2): 178–99.

——. 1998. *Freaks Talk Back: Tabloid Talk Shows and Sexual Nonconformity*. Chicago: University of Chicago Press.

——. 2000, April 17. "Whose Millennium March?" *The Nation*, pp. 16–20.

Gauntlett, David, ed. 2000. *Web.studies: Rewiring Media Studies for the Digital Age*. New York: Oxford University Press.

"Gay Web Portal Buys Publisher." 2000. Online. www.USAToday.com (March 22).

Hargittai, Eszter. 2000. "Open Portals or Closed Gates? Channeling Content on the World Wide Web." *Poetics* 27(4): 233–54.

Hillis, Ken. 1999. *Digital Sensations: Space, Identity and Embodiment in Virtual Reality*. Minneapolis: University of Minnesota Press.

Jones, Steven. 1997. *Virtual Culture: Identity and Community in Cyberspace*. London: Sage.

Kellner, Douglas. 1990. *Television and the Crisis of Democracy*. Boulder, CO: Westview Press.

McChesney, Robert. 2000. "So Much for the Magic of Technology and the Free Market: The World Wide Web and the Corporate Media System," pp. 5–35 in *The World Wide Web and Contemporary Cultural Theory*, ed. A. Herman and T. Swiss. New York: Routledge.

Meland, Marius. 2001. "The Internet and Sexual Orientation." Online. www.eMarketer.com (March 8).

Montopoli, Brian. 2002, April 12. "Blade Runners: Under Corporate Ownership, the Washington Blade Has Lost Its Edge." *Washington City Paper*. [Page numbers unavailable.]

Napoli, Philip M. 1999, Autumn. "Deconstructing the Diversity Principle." *Journal of Communication*, 7–34.

Online Partners. 2001a. "About Online Partners and Gay.com." Online. http://www.onlinepartners.com/about.html.

——. 2001b. "Gay.com and PlanetOut.com Visitors Name Hyundai's 'Boy Toy' The 'Gayest Commercial of All Time.'" Press release, June 5.

——. 2001c. "PlanetOut Partners and Modalis Announce World Research Panel of Lesbian, Gay, Bisexual and Transgender People." Press release, May 2.

——. 2001d. "PlanetOut.com and Gay.com Sponsor Wotapalava Gay Pop Music Festival." Press release, May 30.

Patelis, Korinna. 1999. "The Political Economy of the Internet," in *Media Organisations in Society*, ed. J. Curran. London: Edward Arnold. [Page numbers unavailable.]

PlanetOut. 2000. "Two Largest Online Businesses Serving Gay and Lesbian Market Will Form Global Media and Services Company." Press release, November 16.

——. 2001a. "About PlanetOut." Online.

——. 2001b. "Advertise on PlanetOut." Online. http://www.planetpartners.com/sales.html.

——. 2001c. "PlanetOut Partners Receives $8.2 Million in Initial Series D Funding Round." Press release, May 10.

——. 2001d. "PlanetOut Raises More than $10 Million from Premier Investors." Press release, September 21.

PlanetOut Partners. 2001. "PlanetOut Partners and Rivendell Marketing Create Gay Media Alliance." Press release, October 17.

Poletta, Francesca, and James M. Jasper. 2001. "Collective Identity and Social Movements." *Annual Review of Sociology* 27: 283–305.

Schuman, Susan. 2000. Author interview, San Francisco, June 15.

Scott, Henry E. 2001. "An Urgent Appeal to Lesbian and Gay Community Leaders." E-mail letter, January.

Seidman, Steven. 1993. "Identity Politics in a 'Postmodern' Gay Culture: Some Historical and Conceptual Notes," pp. 105–142 in *Fear of a Queer Planet*, ed. M. Warner. Minneapolis: University of Minnesota Press.

Selvin, Lowell. 2000. "Gay.com Strives to Make an Impact, In Addition to Profit." Online. www.Gay.com (August 24).

Smith, Megan. 2000. Author interview, New York, June 2.

Smith, Rhonda. 2002, March 1. "PlanetOut Strives to Reach Profit Goals." *Washington Blade*. [Page numbers unavailable.]

Stern, Christopher. 2001, May 17. "The Blade Turns a Page." *Washington Post*.

Stille, Alexander. 2001, June 2. "Adding Up the Costs of Cyberdemocracy." *New York Times*, B9, B11.

Streitmatter, Rodger. 1995. *Unspeakable: The Rise of the Gay and Lesbian Press in America*. Boston: Faber and Faber.

Sunstein, Cass. 2001. *Republic.com*. Princeton, NJ: Princeton University Press.

Swidler, Ann. 1986. "Culture in Action: Symbols and Strategies." *American Sociological Review* 51: 273–86.

Taylor, Verta, and Nancy Whittier. 1992. "Collective Identity in Social Movement Communities: Lesbian Feminist Mobilization," pp. 104–29 in *Frontiers in Social Movement Theory*, ed. A. Morris and C. Mueller. New Haven, CT: Yale University Press.

Turkle, Sherry. 1997. *Life on the Screen: Identity in the Age of The Internet*. New York: Touchstone.

Valocchi, Steve. 1999. "The Class-Inflected Nature of Gay Identity." *Social Problems* 42(2): 207–24.

Wakeford, Nina. 2000. "Cyberqueer," pp. 403–415 in *The Cybercultures Reader*, ed. D. Bell and B. M. Kennedy. London: Routledge.

Williams, Rhys. 1995. "Constructing the Public Good: Cultural Resources and Social Movements." *Social Problems* 42(1): 121–44.

Woodland, Randal. 2000. "Queer Spaces, Modem Boys and Pagan Statues," pp. 416–31 in *The Cybercultures Reader*, ed. D. Bell and B. M. Kennedy. London: Routledge.

EPILOGUE

Current Directions and Future Questions

David Silver

Over the last ten years, the field of Internet studies has become, to borrow a term from economists, a "growth industry." Stretching back to the early- and mid-1990s with the work of such thinkers as Julian Dibbell (1993), Howard Rheingold (1993), and Sherry Turkle (1995), and gaining steam during the mid- to late-1990s with the influential anthologies edited by Steve Jones (1995, 1997), the field of Internet studies is currently undergoing institutionalization with the help of the series publication of *New Media and Society* in 1999, the first conference of the Association of Internet Researchers in 2000, and select universities offering degree programs on the topic.

Within this emerging field of study, a healthy amount of energy has been spent addressing questions surrounding the intersections between the Internet and politics. Indeed, with books like Wayne Rash's *Politics on the Net* (1997), Richard Davis's *The Web of Politics* (1999), Elaine Kamarck and Joseph Nye's *Democracy.com?* (1999), and Anthony Wilhelm's *Democracy in a Digital Age* (2000), the topic of "online politics" is rapidly becoming one of the key topics of Internet studies. However, as interesting and important as the aforementioned works are, for the most part they approach their site of inquiry through the eyes of established institutions, focusing more on what politicians and political parties are doing *with the Net* and less on what citizens and activists are doing *on the Net*.

It is for this reason that the contributions that make up *Cyberactivism* supply fresh thoughts and new questions to an already healthy field of study. Interested in issues of politics, democracy, and social capital, the contributors to *Cyberactivism* are especially focused on questions of active engagement, or engaged activism, within informational and/or information-based environments. When approached as "texts," such environments prove tricky; they are most often decentered, fluid, temporary, and subject to constant change brought about by historical, economic, political, and technological developments. For this reason, the interdisciplinary team of contributors, representing no less than eleven different disciplines and fields, is a model for which the editors should be praised and followed.

As with all new fields of inquiry, we are well advised to begin by defining what we mean by the term "cyberactivism." What are its key characteristics? How does it relate to more traditional and less "cyberesque" forms of activism? What are its historical antecedents? Second, we must experiment with and develop (and continue to experiment and develop) a set of methods and theories with which to approach our topic. Is it possible to apply existing methods and concepts, and, if so, do they require specific tweaking and retooling? Is it advantageous to develop new methods, and, if so, are there particular scenarios most suitable to strategically implement them? Third, countering a problematic strain of techno-boosterism—or what Korinna Patelis (1999) calls "Internet-philia"— that has informed much existing scholarship about the Internet, we should be careful to explore not only the potentials but also the pitfalls of cyberactivism. In what ways, we must ask, can the same informational strategies used for progressive social change serve to reinforce existing inequalities and create new ones? How will the race to commercialize cyberspace alter and affect online social and political engagement?

As Jonathan Sterne (1999) warns us, Internet researchers must resist fetishizing our object of inquiry. The Net should not and cannot be approached as a monolithic and static entity. Instead, it is a complicated and evolving technology offering a host of diverse uses to a spectrum of diverse users within a complex sphere of social, cultural, political, and economic contexts, a point understood and supported by many of the contributors to *Cyberactivism*. For example, as Sandor Vegh's chapter shows, organizers of anti-globalization protests against the World Bank incorporated the Internet in a number of ways, including internal and external communications, organization, and mobilization. The various cyberactivist tactics ranged from the mundane (organizing transportation, accommodations, and necessary provisions for protests sites) to more edgy and experimental, such as staging "virtual sit-ins," hacking into Web sites and defacing them, and Internet-transmitted, laser-projected messaging to world leaders. Tracing the creativity of the young and tech-savvy movement, Vegh places the various tactics into three useful categories: awareness/advocacy; organization/mobilization; and action/reaction.

Similarly, Laura J. Gurak and John Logie offer a number of characteristics of both recent and contemporary cyberactivism. These characteristics include speed and reach (positive) and problems with fact checking, credibility, and power structures (negative), which, when applied to their two case studies, provide interesting insight into the effectiveness of online protests. As Gurak and Logie note, sometimes benefits are gained with accompanying expenses: Speed is good but not when adequate forms of vetting and verification are lost; massive reach can be effective but not necessarily when participants are only mildly engaged in the movement.

Dorothy Kidd traces the history and landscape of the influential Seattle chapter of the Independent Media Center (IMC). Although the Seattle IMC uses the Net in creative ways, it is not the only activist tool they use. As Kidd notes, "[t]he four-hundred-strong crew also used all the old and new media, from pens to laptops, inexpensive audio-tape and camcorders to the latest in digital technologies" (61). The "product," therefore, was not only an interactive Web site, but also a suite of distributive multimedia, including a daily newspaper, a daily radio program, and educational videotapes. While Kidd offers many reasons for this, including economic and access issues, one of the most interesting ones to me is the way in which a multimedia ("old" and "new") approach allowed for a richer collaboration; Kidd notes, "[t]his high level of cooperation helped to break down, if not eliminate, the old craft and territorial divisions," (61). Similar to the collaborations between the Zapatistas and the Electronic Disturbance Theatre as noted by Sandor Vegh in this volume, the Seattle IMC combined art and activism, artists and activists to foster a truly unique and engaging form of cyberactivism.

A final characteristic of cyberactivism provided by the contributors is historical. Larry Elin's case study of Zeke Spier shows that this historical element can be deeply personal. Yet it can also take on larger proportions. As noted by Gurak and Logie, the interface of cyberactivism has changed over time as a result of technological developments, from the text-based protests around Lotus MarketPlace and the Clipper chip to more recent Web-based protests around Yahoo!'s shady attempt to appropriate the content of GeoCities. Another historical dimension is one of institutional histories, as seen in Kidd's brief overview of the development of the Seattle IMC. Finally, it is useful when we place cyberactivism as a whole into historical context, as Kidd does when she places it within the less commercial and more socially engaged history of online interaction, stretching back to novel projects like Berkeley Community Memory, Peacenet, and the Association for Progressive Communications.

If Part I of *Cyberactivism* focuses on the topic's defining characteristics, Part II provides us with a number of ways to approach the field of study. Neither trying to nor accidentally succeeding in offering an exhaustive list of approaches, the contributors to Part II suggest four productive methods for studying cyberactivism, and their findings speak to the utility of their methods. These approaches use Habermasian notions of the public sphere, concepts of collective identity, social network analysis, and user identity construction.

Lee Salter asks whether and in what ways the Internet can provide citizens with a healthy public sphere, as articulated by Jürgen Habermas in *The Structural Transformation of the Public Sphere* (1989), as well as in *The Theory of Communicative Action* (1987) and *Between Facts and Norms* (1996). Like Kidd, Salter understands that the Net is, among many things, a historical construction and uses the development of the Internet in general and the formation of the Association for Progressive Communications in particular as his primary focal points. In both cases, he notes the importance and utility of what has become known as "participatory design" (Schuler and Namioka 1993) or, alternatively, "community informatics" (Gurstein 2000; Keeble and Loader 2001), suggesting, for example, that the Net's early stages of development benefitted greatly from open and free discussion regarding technical problems and solutions, as well as an open-source distribution of code and software.

That said, Salter is cautious not to equate cyberspace in its current manifestations with the public sphere as idealized by Habermas, a problematic conflation too often unexplored by early Internet studies scholarship. For Salter, the problem rests chiefly in what Habermas details as the requisite conditions for successful communicative action. Echoing Gurak and Logie, Salter notes that accuracy of information, a shared cultural background, and sincerity cannot be assumed on the Net.

The differences (and similarities) between online and offline collective action is elaborated in Michael D. Ayers's essay. Although Ayers's major findings revolve around concepts of collective identity, his useful research design merits discussion. Unlike so much scholarship on how individuals and groups use the Net, Ayers's work is comparative; he selects two very different sites of inquiry: NOW Village, a primarily online organization, and Womanspace, a primarily face-to-face one. This comparative study is made richer by Ayers's learning firsthand from the group members; he replaces media press clippings with in-depth interviews, an approach I applaud for its willingness to go beyond media-friendly sound bites and venture firsthand into the culture and its participants. Ayers's work shows that unlike some of the earlier and now nearly canonized works on virtual communities (Baym 1999; Correll 1995; Dibbell 1993; Rheingold 1993), the members of NOW Village appear to be enjoying very few shared experiences.

While the section's first two chapters employ more social scientific methods, the last two, one by Maria Garrido and Alexander Halavais and the other by Wyatt Galusky, offer more technological methods, approaches that contribute significantly to the emerging foundation of what has and should be called "interface analysis." Interface analysis refers to the recent move to study the interface as a site of culture, as a significant player in the kinds of interactions made possible, promoted, and thwarted by various Web designers. While "classic" Internet studies scholars have focused primarily on the interactions that take place within online environments, recent scholarship considers the role that design plays with respect to such interactions (Kolko 2000; Lessig 1999; Silver 2000). For example, Garrido and Halavais are interested not only in the Zapatista movement in cyberspace, but also in "the ties, roles, and strategic alliances that have been built within and around the movement worldwide" (166). Such an exploration lends itself to social network analysis, a method by which patterns of exchange and relationships among groups are uncovered, mapped, and examined.

Their scope is both deep and daunting; using a customized Web crawler, one hundred thousand pages were crawled, resulting in several million hyperlinks stemming from 392 domains. What they find is fascinating–that the Zapatista movement in cyberspace represents a cross-national node of collective solidarity. At times strategic, at times loosely if not weakly affiliated, this network presents activists and scholars alike with a model of what it takes to build an online global civil society. Fittingly, Garrido and Halavais conclude with a challenge, one that can, should, and must be reckoned with: "If we are to discuss these networks as social fact, as something being built through discourse and action, we must do more than acknowledge their presence" (182). This is a large and important task. Indeed, we must begin to approach individual nodes on the Net as not free-floating spaces of production, but rather as a massively connected and interconnected nodes of interaction. There are a few islands on the Net; it's time we begin studying the bridges.

Like Garrido and Halavais, Galusky's interests rest within the intersections between civic engagement and the interface, yet focus on interfacial constructions of individual users/citizens. Examining the environmental anti-toxins movement's use of the Net in general and the activist Web site www.scorecard.org in particular, Galusky notes the problematic ways in which citizen activists are constructed by and in relation to online environments.

For Galusky, www.scorecard.org offers users "consumptive empowerment," without the opportunity to voice their own insights, experiences, and local expertise. While Galusky notes the site's ability to provide users with an abundance of *information*, he warns us against conflating information with *empowerment*, a conflation that may prevent more committed involvement. Indeed, as Galusky and many other of the volume's contributors note, the best kind of online activism engages users as active agents of change rather than passive consumers of information.

The three chapters that comprise Part III rest somewhere between cyberhype and cybergripe. Hopeful that new media can contribute significantly and creatively toward progressive movements yet unwilling to argue that such media contain within them some inherent seed for change, the three chapters travel somewhere in between the promise and peril of current cyberactivism. Future cyberactivists should surely take heed.

Joanne Lebert provides an interesting case study of Amnesty International's use of information and communication technologies to illustrate some of the shortcomings of cyberactivism. While acknowledging their utility, especially when used in concert with more established tools, Lebert also notes some of their limitations, which include massive time and resource commitments, varying degrees of accessibility, and issues of language translation. Further, echoing earlier chapters, Lebert is uncertain about the ways in which offending institutions *receive* cyberactivism, be it in the form of a single e-mail or full-scale e-mail petitions.

Lebert notes the amount of invisible labor cyberactivism requires; cyberactivism demands time, attention, and special technical resources and skills to be effective. A few examples should suffice: Adhering to the principles of participatory design may be an effective way to build democratic action, but it requires massive voluntary effort; building online alliances, like those found in the Zapatista movement, may be useful, but they also generate additional responsibilities and time-consuming monitoring activities; and while making available materials regarding aspects of injustice is crucial, it requires around-the-clock support, references, and updating. Fighting the power has never been easy, and although the Net may make some elements more convenient and effective, it brings with it a new host of concerns and responsibilities, some of which are hard to see.

It is also important to note that Lebert's analysis comes *from within*. As I note later, some of the best Internet studies scholarship takes on elements of ethnography; scholars become participants and observers. Lebert's insider role—she serves as the Urgent Action Coordinator for the English-speaking Canadian division of Amnesty International—blends well with her training in social anthropology, producing a first-hand look into the movement's offline and online actions. These are the kinds of insider case studies that we can and should be producing.

It is puzzling to note that the topic of commercialization, by far the most dominating aspect of cyberculture during the last few years, is by far the most ignored direction of inquiry within Internet studies. In this light, Steven McLaine and Joshua Gamson's chapters are important correctives: they remind us to think critically about cyberactivism in an age of .com capitalism, demanding that we place our topic of study within larger economic and consumer contexts.

"Profit and community make curious bedfellows," writes McLaine in his analysis of three Web sites for "ethnic affinity groups." While acknowledging the (limited) utility of sites like AsianAvenue.com, BlackPlanet.com, and MiGente.com, including sporadic yet somewhat successful attempts to mobilize social and consumer protests, we must beware of cookie-cutter sites that use "difference" as a market strategy. The problematic line between commerce and community is exposed upon analyzing the kinds of advertising the site hosts. As McLaine notes, online banking and credit card advertisements adorn BlackPlanet.com, despite repeated criticism regarding the loan practices of these banks within African American communities. Echoing the work of Lisa Nakamura (1995), McLaine notes that many of the users' handles—FineAssAznGurl, ThugRidah_80, Fili_Dragon, and hoocheemama, to name a few examples from AsianAvenue.com—assume the same kind of cultural (not to mention racist and sexist) stereotypes consistently found in traditional media.

Gamson is equally suspicious of for-profit Web sites promising to empower and/or organize groups linked by a "marginal" identity. He begins by critiquing recent scholarship linking queer theory with the Net's supposed ability to make possible fluid identities and performative genders. Acknowledging such possibilities, Gamson focuses instead on the important structural elements brought about by the convergence between the Net, the (until recently) massive influx of capital investment, and consumerism. Like McLaine, Gamson is concerned with what he perceives as a particularly nasty strain of marketing-as-liberation logic taking place within gay community sites; he notes that "[s]erving the community and penetrating the market are one and the same" (265). And the result? With the help of massive investments and corporate buyouts, so-called marginal media sources get appropriated by larger, more mainstream media entities and produce materials friendly to investors and minimally interesting to their users.

By assembling *Cyberactivism*, Martha McCaughey and Michael D. Ayers have provided activists and scholars alike a blueprint for future political engagement and academic research. We learn about what cyberactivism is and what it can be, its potentials, promises, and perils. *Cyberactivism* allows us to better understand multiple forms of activist movements and networks while appreciating the various directions activists, artists, and technologists can and should take in the future.

While the volume's contributors offer a number of fruitful directions future studies of online activism can take, I wish to conclude by briefly suggesting a few more. To date, the majority of historically based Internet studies scholarship has fallen within two camps: the first traces the development of the Internet (Abbate 1999; Hafner and Lyon 1996; Hauben and Hauben 1997); the second places the Internet within more broad histories of communication technologies (Peters 1999; Sconce 2000; Standage 1998; Wertheim 1999). While useful, we must work diligently and creatively to establish a third camp, one that historically contextualizes our particular case studies, as do many of the contributors. Whether online or offline, social and political movements develop over time, time that invariably brings with it new dilemmas, new directions, new problems, and new solutions. As contributors to the young field of Internet studies, scholars studying cyberactivism stand to gain much from as many historical case studies as possible.

These histories can tell us what happened, what could have happened, and what should have happened. By historically contextualizing cyberactivism, we better understand it as a set of fluid, changing, and *changeable* ebbs and flows.

Historical study is difficult work, and studying the histories of online activist movements is no exception. Perhaps that is one of the reasons that Internet studies scholars have for the most part failed at historically contextualizing their sites of study. Yet it bears repeating: For the study of cyberactivism to flourish, we must take an opposite approach and begin by historically situating our studies. Among the questions we can and should ask are: Who were the principal project partners and what were their respective roles and contributions? What were their initial goals and visions for the projects? Were the goals and visions largely commercial or communitarian, or a combination of both? At what stage, if any, were future users invited to brainstorm, comment on, and contribute to the building of the networks?

Second, we should direct more attention to rhetorical components of cyberactivism. As Laura J. Gurak (1997), Lisa Nakamura (2000), and Barbara Warnick (1999) show us, cyberculture, like all forms of culture, is in part rhetorically constructed; the stories we hear and share about the Net influence and inform the ways in which we approach the Net. In other words, our understanding of online movements like those organized by the Zapatistas and Amnesty International should incorporate stories spun about such movements. We must analyze the stories in the popular press, press releases issued to feed or challenge such stories, and the framing that takes place when such stories are put together.

Indeed, while many of the volume's contributors applaud the efforts taken by those involved in the various projects discussed, media coverage of the same projects often tell another story. Future studies should examine the discourses surrounding online movements like those organized by the Zapatistas and Amnesty International. Further, we must also be aware that scholars' framing of these moments will impact the movements as well. Perhaps a future goal to work toward would be the establishment of a scholars-activists network, a (virtual) space where the two camps can come together, share stories and strategies, and work for a better understanding and more effective implementation of online activism.

Third, taking our cue from existing ethnographic-based scholarship in Internet studies (Baym 2000; Danet 2001; Hine 2000; Markham 1998), we must talk with, listen to, and learn from the many players of cyberactivism. Through surveys, in-depth interviews, and participant-observation, we should collect our stories firsthand: from activists and artists, from offline site coordinators and online site designers, from institutions targeting the social movement, and from institutions targeted by the social movement. Indeed, online activism is not and cannot be contained within a single party line or origins myth. To better understand our subject, we must engage firsthand with its multiple and diverse players.

Fourth, we must expand upon the interfacial analyses offered by many of the contributors. The interfaces upon which cyberactivism are played out are as varied and complex as the tactics, goals, and ideologies of its organizers. As such, we must get our hands dirty with design and become "digital archaeologists" with code as our test site. For example, whether it is the members of the rock group Radiohead using black text on black background to hide subversive messages on their Web site or hackers setting up dummy "404: File Not Found" sites to serve as portals for pirated software, things are not always as they appear on the screen. For that reason, special attention must be spent on code, ranging from line-to-line analysis of html to becoming more knowledgeable about the fine tunings of "cookies" and other invisible elements of the Net.

Of course, our test site is in flux to say the least: ftp becomes gopher becomes the Web; html is replaced by cascading style sheets which is replaced by xml. And for that reason, we must continue our interdisciplinary approaches; much can and should be learned from scholarship in design, human-computer interaction, and visual culture, to name only a few. We also can and should talk and listen to those responsible for the code; the Web designers and teams behind activist sites often have as much to say as those behind the bullhorns.

And finally, while the collected chapters of *Cyberactivism* do an outstanding job at exploring, documenting, and analyzing online activism *on the left*, we will be doing ourselves a disservice if such an analysis is done at the expense of a broader political spectrum of activism. Indeed, while anti-WTO organizers develop sophisticated forms of online culture jamming, the corporate right are no novices, as revealed in such campaigns as oil companies producing commercials about environmentalism or the kind of spot-on press feeds delivered by the White House during these post–September 11 times. For that reason, more analysis of online activism *on the right* is needed.

Although it would be untrue to say that the field of Internet studies is inherently leftist, it is fair to say that much of its scholarship is informed by socially and politically progressive ideas and ideologies, as witnessed in the contributions to *Cyberactivism*. That said, scholars and activists on the left should be more aware of the activities going down on the other side. Of course, it would be interesting to see a similar collection of analyses charting online activism *from the right, by the right*, especially as such activism has gained considerable momentum since September 11, 2001.

Both the scholarship of and the activism documented in *Cyberactivism* are the first flowers of their kind. May hundreds more bloom.

References

Abbate, Janet. 1999. *Inventing the Internet*. Cambridge, MA: MIT Press.

Baym, Nancy K. 1999. *Tune In, Log On: Soaps, Fandom, and Online Community*. Thousand Oaks, CA: Sage.

Correll, Shelley. 1995. "The Ethnography of an Electronic Bar: The Lesbian Cafe." *Journal of Contemporary Ethnography* 24(3): 270–98.

Danet, Brenda. 2001. *Cyberpl@y: Communicating Online*. Oxford: Berg Publishers.

Davis, Richard. 1999. *The Web of Politics: The Internet's Impact on the American Political System*. Oxford: Oxford University Press.

Dibbell, Julian. 1993, December 23. "A Rape in Cyberspace; or How an Evil Clown, a Haitian Trickster Spirit, Two Wizards, and a Cast of Dozens Turned a Database into a Society." *The Village Voice*, 36–42.

Gurak, Laura J. 1997. *Persuasion and Privacy in Cyberspace: The Online Protests over Lotus MarketPlace and the Clipper Chip*. New Haven, CT: Yale University Press.

Gurstein, Michael, ed. 2000. *Community Informatics: Enabling Communities with Information and Communications Technologies*. Hershey, PA: Idea Group Publishing.

Hafner, Katie, and Matthew Lyon. 1996. *Where Wizards Stay Up Late: The Origins of the Internet*. New York: Simon and Schuster.

Hauben, Michael, and Ronda Hauben. 1997. *Netizens: On the History and Impact of Usenet and the Internet*. Los Alamitos, CA: IEEE Computer Society Press.

Hine, Christine. 2000. *Virtual Ethnography*. London: Sage.

Jones, Steve, ed. 1995. *CyberSociety: Computer-Mediated Communication and Community*. Thousand Oaks, CA: Sage.

——. 1997. *Virtual Culture: Identity and Communication in Cybersociety*. Thousand Oaks, CA: Sage.

Kamarck, Elaine, and Joseph Nye. 1999. *Democracy.com? Governance in a Networked World.* New York: Hollis Publishing.

Keeble, Leigh, and Loader, Brian, eds. 2001. *Community Informatics: Shaping Computer-Mediated Social Networks*. London and New York: Routledge.

Kolko, Beth. 2000. "Erasing @race: Going White in the (Inter)Face," pp. 213–32 in *Race in Cyberspace*, ed. Beth Kolko, Lisa Nakamura, and Gilbert Rodman. New York: Routledge.

Lessig, Lawrence. 1999. *Code and Other Laws of Cyberspace*. New York: Basic Books.

Markham, Annette N. 1998. *Life Online: Researching Real Experience in Virtual Space.* Walnut Creek, CA: AltaMira Press.

Nakamura, Lisa. 1995. "Race in/for Cyberspace: Identity Tourism and Racial Passing on the Internet." pp. 181–193 in *Works and Days* 25/26, 13(1–2).

——. 2000. "Where Do You Want to Go Today?: Cybernetic Tourism, the Internet, and Transnationality," pp. 15–26 in *Race in Cyberspace*, ed. Beth Kolko, Lisa Nakamura, and Gilbert Rodman. New York: Routledge.

Patelis, Korinna. 1999. "The Political Economy of the Internet," pp. in *Media Organisations in Society*, ed. J. Curran. London: Edward Arnold.

Peters, John Durham. 1999. *Speaking into the Air: A History of Idea of Communication*. Chicago: University of Chicago Press.

Rash, Wayne Jr. 1997. *Politics on the Net: Wiring the Political Process*. New York: W. H. Freeman.

Rheingold, Howard. 1993. *The Virtual Community: Homesteading on the Electronic Frontier*. Reading, MA: Addison Wesley.

Schuler, Douglas, and Aki Namioka, eds. 1993. *Participatory Design: Principles and Practices*. Hillsdale, NJ: Lawrence Erlbaum Associates.

Sconce, Jeffrey. 2000. *Haunted Media: Electronic Presence from Telegraphy to Television*. Durham, NC, and London: Duke University Press.

Silver, David. 2000. "Margins in the Wires: Looking for Race, Gender, and Sexuality in the Blacksburg Electronic Village," pp. 133–50 in *Race In Cyberspace*, ed. Beth Kolko, Lisa Nakamura, and Gilbert Rodman. New York: Routledge.

Standage, Tom. 1998. *The Victorian Internet: The Remarkable Story of the Telegraph and the Nineteenth Century's On-Line Pioneers.* New York: Berkley Publishing Group.

Sterne, Jonathan. 1998. "Thinking the Internet: Cultural Studies vs. The Millennium," pp. 257–88 in *Doing Internet Research*, ed. Steve Jones. Thousand Oaks: Sage.

Turkle, Sherry. 1995. *Life on the Screen: Identity in the Age of the Internet.* New York: Simon and Schuster.

Warnick, Barbara. 1999. "Masculinizing the Feminine: Inviting Women Online ca. 1997." *Critical Studies in Mass Communication* 16: 1–19.

Wertheim, Margaret. 1999. *The Pearly Gates of Cyberspace: A History of Space from Dante to the Internet.* New York and London: W.W. Norton.

Wilhelm, Anthony. 2000. *Democracy in a Digital Age: Challenges to Political Life in Cyberspace.* New York: Routledge.

ABOUT THE CONTRIBUTORS

Michael D. Ayers is a graduate student in sociology in the Graduate Faculty of Political and Social Science at New School University in New York City. He received his M.S. in sociology from Virginia Tech. In addition to his work in cultural studies of the Internet and social change, he is conducting an ethnographic research project at a New York City art collective examining the role of space in relation to group identity.

Larry Elin teaches writing, production, media business, and multimedia at the S. I. Newhouse School of Public Communications at Syracuse University. Before joining the faculty in January 1998, he had spent twenty-five years as a creative executive in interactive multimedia, interactive television, feature films, animation and special effects, television graphics, and advertising. Mr. Elin is widely known as a pioneer in 3-D computer graphics and in interactive multimedia. He has received numerous industry awards, including a Clio in 1983, a Grand Prix (graphics) in 1984, a Gold Medal at the New York International Film Festival in 1984, a silver from the Broadcast Designers Association in 1993, and a nomination for the Presidential Leadership award from

GTE. He is the author of *Designing and Developing Multimedia*, published by Allyn and Bacon, and coauthor of the book *Hype, Humility, Hope: The Real Promise of Virtual Political Communities*, to be published in 2002. His research includes the design and development of an interactive multimedia teaching tool to instruct young African American females on strategies to avoid contracting HIV. He is cochair of the Media and American Democracy institute at the Newhouse School, Syracuse University. He is a member of the Academy of Interactive Arts and Sciences and the Broadcast Education Association.

Wyatt Galusky is a Ph.D. candidate in science and technology studies at Virginia Tech. He has published research on environmentalism and the role of politics and the public sphere in *Organization and Environment* and *The Bulletin of Science, Technology and Society.* He received his B.A. in psychology from the Texas A&M (1993) and his M.A. in philosophy/environmental ethics from the University of North Texas (1997). His current research focuses on environmental ethics, risk, and risk communication.

Joshua Gamson is an associate professor of sociology at Yale University, author of *Freaks Talk Back: Tabloid Talk Shows and Sexual Nonconformity* (1998), *Claims to Fame: Celebrity in Contemporary America* (1994), and a participating author of *Ethnography Unbound: Power and Resistance in the Modern Metropolis* (1991). *Freaks Talk Back* won the Society for Cinema Studies Kovacs Award and the American Sociological Association Sociology of Culture Book Award and was selected as one of *The Voice Literary Supplement*'s twenty-five favorite books of 1998. His research and teaching focus on the sociology of culture, with an emphasis on contemporary Western commercial culture and mass media; social movements, especially cultural aspects of contemporary movements; participant-observation methodology and techniques, particularly as applied in urban settings; and the history, theory, and sociology of sexuality. Recent published work includes studies of sex scandal discourse in the United States (*Social Problems*), the political pitfalls in the pursuit of media visibility (*Sexualities*), exclusion processes in sex/gender movements (*Gender and Society*), organizational aspects of collective identity construction (*Sociological Forum*), and dilemmas in identity-based movements (*Social Problems*). He has also been published in *The Nation*, *The American Prospect*, *Tikkun*, *The Utne Reader*, and *The New Yorker*. Prof. Gamson received his Ph.D. in sociology from the University of California at Berkeley.

Maria Garrido is a doctoral student in the Department of Communication at the University of Washington. Her research explores the role of information technology in fostering economic development in low-income communities, particularly in Latin America, concentrating on alternative solutions to the complex nature of the digital divide. She is also researching the way in which social movements make use of new media as a tool for mobilization and to create networks of support on a global scale.

Laura J. Gurak is an associate professor in the Department of Rhetoric at the University of Minnesota and the director of the Internet Studies Center. She is the author of *Cyberliteracy: Navigating the Internet with Awareness* (2001) and *Persuasion and Privacy in Cyberspace: The Online Protests Over LotusMarketplace and the Clipper Chip* (1997). She has published articles in *Technical Communication*, *Computers and Composition*, and *Rhetoric Review*. Her current research interests include rhetoric of technology, intellectual property, and Internet studies. She received her Ph.D. from Rensselaer Polytechnic Institute in St. Troy, New York (1994).

Alexander Halavais studies the ways in which new communication technologies facilitate large-scale interaction. By examining the hyperlinked structure of the World Wide Web and other networked systems of communication, he aims to describe how social change and creative solutions occur within large, non-hierarchical groups. His interests also include the structures of globalization, especially emergent global communities and transnational urban networks. Prior to joining the faculty of the School of Informatics at the University at Buffalo/SUNY in 2001, Halavais spent several years teaching in Japanese public schools, worked in communications positions in the public and private sector, and acted as the Research Director for the New Media Research Lab at the University of Washington. In 2001 he received his Ph.D. in Communications at the University of Washington.

Dorothy Kidd is an assistant professor in the Department of Media Studies at the University of San Francisco. She was the coeditor for a special issue of *Peace Review* on the topic of social justice movements and the Internet. She has published articles in *MediaFile* and *Whole Earth Review*. She received her B.A. in radio and TV arts from Ryerson Polytechnical University in Toronto, Ontario, Canada, where she was the recipient of the Ted Pope Memorial Award for Humanities (1979). She received her M.A. and Ph.D. in communications from Simon Fraser University in Burnaby, British Columbia, Canada (1990, 1998). She was awarded the Social Sciences Humanities Research Council Award for her dissertation "The Media Enclosures and Communications Commons: Field Work in Central America and Vancouver."

Joanne Lebert is a Ph.D. candidate in social anthropology at York University in Toronto, Ontario, Canada. She is also the Urgent Action Coordinator for the English-speaking Canadian division of Amnesty International. She is the author of *Information and Communications Technologies and Human Rights Advocacy: The Case of Amnesty International in Civil Society in the Information Age: NGO's, Coalitions and Relationships* (2002). Along with cyberactivism, her research focuses on the construction and local interpretation of human rights and national reconciliation, identity politics, and sociopolitical conflict, especially as these pertain to Angola and Southern Africa. She received an honors B.A. from Toronto University (1996), a graduate diploma in refugee and migration studies from York University (1999), as well as an M.A. in social anthropology from York University (1999). Joanne was also a visiting study fellow in the Refugee Studies Program at the University of Oxford (1997–1998).

John Logie is an assistant professor in the Department of Rhetoric at the University of Minnesota as well as the codirector of the Internet Studies Center. He has recently published articles in *Rhetoric Society*

Quarterly (2001), *Technical Communication Quarterly* (1998), *Computers and Composition* (1998), and the *Journal of Technical Writing and Communication* (1996). He received a B.A. in honors English and creative writing from the University of Michigan, an M.A. in English from the University of Illinois at Chicago, and a Ph.D. in English from Pennsylvania State University. His current research focuses on questions of intellectual property and ownership and their intersections with understandings of rhetorical and literary invention. He is currently working on a book-length project on the music file-swapping service Napster, and also works on rhetorical delivery, especially as it relates to the Internet.

Martha McCaughey is an associate professor and director of women's studies at Virginia Tech. She is the author of *Real Knockouts: The Physical Feminism of Women's Self Defense* (1997) and coeditor of *Reel Knockouts: Violent Women in the Movies* (University of Texas Press, 2001). She has published articles on gender, aggression, the body, and social change. She is also lead author of *The Mayfield Quick-View Guide to the Internet for Students of Women's Studies* (2000) and of the essay "Cybergrrrl Education and Virtual Feminism: Using the Internet to Teach Introductory Women's Studies" (1999).

Steven McLaine is a research fellow at the Center for the Study of Voluntary Organizations and Service, a nonprofit research center affiliated with the Georgetown Public Policy Institute at Georgetown University. His most recent work, "Low Income and Minority User Satisfaction at Community Technology Centers: A Statistical Analysis of CTC User Data," has been posted on the Community Technology Centers Network (CTCNet) Web site (www.ctcnet.org). Steven received his M.A. in Public Policy from Georgetown University (2001) and his B.A. from the University of Virginia (1993). Additional research interests include the digital divide, technology policy, nonprofit technology, educational technology, and cyberculture.

Lee Salter is currently researching the relation between media, in particular the Internet, and democracy at the University of North London.

He also teaches political theory and institutions at the same institution. He completed his B.A. (honors) at London Guildhall University and his M.Sc. at Birkbeck College, University of London. His research interests also include liberal political theory, democratic theory, mass media, and the thought of Jürgen Habermas. He is an at large member of the Internet Corporation for Assigned Names and Numbers, as well as a member of the Unit for Internet Studies, and the Political Studies Association Media Politics Group.

David Silver is an assistant professor in the School of Communications at the University of Washington and the founder of the Resource Center for Cyberculture Studies. He is the author of chapters on cyberculture in *Race and Cyberspace* (Routledge, 2000), *Web Studies: Rewiring Media Studies for the Digital Age* (2000) and *The Global Village: Dead or Alive?* (1999). He received his Ph.D. in American studies from the University of Maryland (2000). David's current research focuses on the cultural history of e-commerce and the meta-analysis of the field of cyberculture studies.

Sandor Vegh is a Ph.D. candidate and Fulbright Scholar in the Department of American studies at the University of Maryland. Sandor received a B.S. in mathematics, a B.S. in computer science, and a M.A. in English (American studies) from the University of Debrecen, Hungary (1996). He has published work on the Internet's impact on democracy in *InterSections* as well as research on the American media in the Hungarian journal *Mediakonyv*. His current research interests focus on the media representations of politically driven hacking, democracy and the Internet, and counterhegemonic uses of the Internet.

INDEX